Multiple Criteria Decision Making

Series Editor

Constantin Zopounidis
School of Production Engineering and Management
Technical University of Crete
Chania
Greece

This book series focuses on the publication of monographs and edited volumes of wide interest for researchers and practitioners interested in the theory of multicriteria analysis and its applications in management and engineering. The book series publishes novel works related to the foundations and the methodological aspects of multicriteria analysis, its applications in different areas in management and engineering, as well as its connections with other quantitative and analytic disciplines. In recent years, multicriteria analysis has been widely used for decision making purposes by institutions and enterprises. Research is also very active in the field, with numerous publications in a wide range of publication outlets and different domains such as operations management, environmental and energy planning, finance and economics, marketing, engineering, and healthcare.

This series has been accepted by Scopus.

Alireza Amirteimoori • Mansour Soufi

Industrial Decision Analysis

Theory and Applications of MCDM Methods

Springer

Alireza Amirteimoori
Faculty of engineering
and Natural Sciences
Istinye University
Istanbul, Türkiye

Mansour Soufi
Department of Industrial Management, Ra.C.
Islamic Azad University
Rasht, Iran

ISSN 2366-0023 ISSN 2366-0031 (electronic)
Multiple Criteria Decision Making
ISBN 978-3-032-24298-3 ISBN 978-3-032-24299-0 (eBook)
https://doi.org/10.1007/978-3-032-24299-0

This Springer imprint is published by the registered company Springer Nature Switzerland AG
The registered company address is: Gewerbestrasse 11, 6330 Cham, Switzerland

Preface

Decision-making in today's complex industrial environments—marked by multiple criteria, uncertainties, and interdependencies among alternatives—poses a significant challenge for managers, engineers, and researchers. Multi-Criteria Decision-Making (MCDM) methods have emerged as powerful tools to systematically analyze multidimensional problems, enabling decision-makers to balance competing objectives such as cost, quality, sustainability, and risk.

This book is designed to establish a robust connection between theoretical foundations and practical applications in the realm of Multi-Criteria Decision-Making. The initial chapters provide a clear exposition of the conceptual and mathematical foundations of key methodologies, including AHP, TOPSIS, DEMATEL, ELECTRE, and PROMETHEE. Subsequent sections present real-world case studies from diverse industries (e.g., supply chain, energy, manufacturing, and technology), demonstrating how these techniques address operational and strategic challenges. A distinctive feature of this work is its integration of advanced approaches—such as fuzzy MCDM, hybridization with machine learning, and decision support under uncertainty—with step-by-step examples, making it equally valuable for academic and professional audiences.

The book targets researchers in industrial engineering, management, economics, and data science, as well as practitioners and consultants seeking quantitative tools to optimize decisions.

Istanbul, Turkey — Alireza Amirteimoori
Rasht, Iran — Mansour Soufi

Competing Interests The authors have no competing interests to declare that are relevant to the content of this manuscript.

Contents

List of Abbreviations

AHP	Analytic Hierarchy Process
ANP	Analytic Network Process
BWM	Best-Worst Method
CI	Consistency Index
CMMS	Computerized Maintenance Management System
COPRAS	Complex Proportional Assessment
CR	Consistency Ratio
CVI	Content Validity Index
CVR	Content Validity Ratio
DANP	DEMATEL-based Analytic Network Process
DEA	Data Envelopment Analysis
DEMATEL	Decision-Making Trial and Evaluation Laboratory
DM	Decision-Maker
DMU	Decision-Making Unit
ELECTRE	ELimination and Choice Translating REality (from French)
F-AHP	Fuzzy Analytic Hierarchy Process
F-Delphi	Fuzzy Delphi Method
F-DEMATEL	Fuzzy Decision-Making Trial and Evaluation Laboratory
F-LLSM	Fuzzy Logarithmic Least Squares Method
F-MADM	Fuzzy Multi-Attribute Decision-Making
F-MCDM	Fuzzy Multi-Criteria Decision-Making
F-PROMETHEE	Fuzzy Preference Ranking Organization Method for Enrichment Evaluations
FANP	Fuzzy Analytic Network Process
GAIA	Graphical Analysis for Interactive Assistance
GRA	Grey Relational Analysis
LFPP	Logarithmic Fuzzy Priority Programming
LINMAP	Linear Programming Technique for Multidimensional Analysis of Preference
LLSM	Logarithmic Least Squares Method

LPP	Linear Priority Programming/Logarithmic Priority Programming (context-dependent)
LSM	Least Squares Method
MADM	Multi-Attribute Decision-Making
MAUT	Multiple Attribute Utility Theory
MCDA	Multi-Criteria Decision Analysis
MCDM	Multi-Criteria Decision-Making
MODM	Multi-Objective Decision-Making
MOLP	Multi-Objective Linear Programming
MOORA	Multi-Objective Optimization on the Basis of Ratio Analysis
PROMETHEE	Preference Ranking Organization METHod for Enrichment Evaluations
RI	Random Index
SAW	Simple Additive Weighting
STEM	Step Method
TOPSIS	Technique for Order of Preference by Similarity to Ideal Solution
TRM	Total Relation Matrix
VIKOR	VlseKriterijumska Optimizacija I Kompromisno Resenje (Serbian)
WASPAS	Weighted Aggregated Sum Product Assessment
WPM	Weighted Product Model
WSM	Weighted Sum Model

List of Figures

List of Tables

Chapter 1
Multi-Objective Decision-Making: Optimizing in a Complex World

Learning Objectives
Upon completion of this chapter, the reader will be able to:

1. **Articulate the nature and significance** of multi-objective decision-making (MODM) in addressing complex, real-world problems characterized by conflicting goals.
2. **Define and explain** the fundamental concepts in MODM, including the *Pareto front*, *Pareto optimality*, and the inherent *trade-offs* between competing objectives.
3. **Categorize** the principal methodologies for solving multi-objective problems based on the timing of preference information incorporation (a priori, interactive, and a posteriori) and **identify** the characteristics of prominent techniques such as *goal programming*, the *weighted sum method*, and the *epsilon-constraint method*.
4. **Outline** the systematic steps involved in formulating and solving an MODM problem, from objective identification and mathematical modeling to solution analysis and final decision-making.
5. **Analyze the broad applicability** of MODM frameworks across diverse domains, including supply chain management, product design, and energy management.
6. **Understand the fundamental principles** of key solution techniques and **apply** basic methods like the *STEM method* and the *global criterion method* to simplified problems.

A. Amirteimoori, M. Soufi, *Industrial Decision Analysis*, Multiple Criteria Decision Making, https://doi.org/10.1007/978-3-032-24299-0_1

1.1 Introduction

Decision-making is an integral part of our daily lives. However, when faced with multiple, often conflicting, objectives, the process becomes significantly more complex. Multi-objective decision-making refers to the systematic process of selecting the best option from a set of alternatives, aiming to simultaneously achieve several distinct goals. These objectives may be aligned or even contradictory.

Multi-objective linear programming (MOLP) problems arise frequently in decision-making processes across various domains, including economics, engineering, transportation, social sciences, and more. In multi-objective optimization, we seek to solve problems involving multiple, often conflicting, objectives that must be optimized simultaneously. Without explicit preference information, we can only assert that the best solution for the decision-maker (DM) lies within the Pareto optimal set of solutions.

In order to incorporate these preferences into the problem and depending on the timing of their application, various techniques have been introduced. These techniques can be broadly categorized into three groups.

The first group of techniques elicits preference information from the decision-maker before solving the problem. Some examples of these techniques include goal programming, lexicographic methods, and bounded objective methods.

The second category includes techniques that involve continuous interaction with the decision-maker and elicit preferences during the solution process. STEM, SWOT, GEOFRION, and ZIONTS techniques fall into this category.

The third category of techniques considers the decision-maker's preference information at the end of the solution process. Parametric methods, epsilon-constraint methods, and multi-objective linear programming methods are examples of these approaches.

In this chapter, we delve into the topic of multi-objective decision-making, a subset of multi-criteria decision-making.

1.2 What Is MODM?

Multi-objective decision-making (MODM) is a pivotal field within management science and operations research. It addresses problems characterized by multiple, often conflicting or complementary, objectives. For instance, a corporation might simultaneously seek to maximize profits, minimize costs, and enhance product quality. Such multifaceted objectives can significantly complicate the decision-making process.

Abdali et al. (2021) stated the significance of multi-criteria decision-making models as follows:

- **Real-world relevance:** Real-world problems are inherently multidimensional, necessitating a holistic approach that considers multiple objectives.

- **Comprehensive optimization:** Rather than focusing on a single optimal solution, MODM identifies a set of efficient solutions where no objective can be improved without compromising another.
- **Facilitated decision-making:** By providing a structured framework, MODM empowers decision-makers to evaluate various alternatives systematically and make informed choices.

1.3 Types of MODM Models

Numerous methods have been developed to address multi-objective decision problems, each suited to specific circumstances. Key approaches include (Legesse Gebre et al., 2021):

- **Goal programming:** Assigns weights or priorities to objectives and seeks to minimize deviations from desired goals.
- **Weighted sum method:** Combines objectives into a single-objective function using weights to represent their relative importance.
- **Ranking methods:** Prioritizes objectives and identifies efficient solutions based on the ranking.
- **Epsilon-constraint method:** Treats one objective as the primary function while others are constrained, allowing for the exploration of a range of efficient solutions.
- **Multi-criteria decision analysis (MCDA):** Suitable for problems with a limited number of alternatives, MCDA helps decision-makers select the best option based on multiple criteria.

1.4 Steps Involved in MODM

Asgarpour (2023) specified the steps involved in multi-objective decision-making as follows:

- **Objective identification and definition:** Clearly define all relevant objectives.
- **Mathematical modeling:** Formulate the problem as a mathematical model, incorporating objectives, constraints, and decision variables.
- **Model solution:** Solve the model using specialized software or numerical methods.
- **Result analysis:** Analyze the obtained solutions to identify the efficient frontier.
- **Decision-making:** Select the most preferred solution based on the decision-maker's preferences and insights.

1.5 Applications of MODM

MODM finds applications in various domains, including (Zahraee et al., 2020):

- **Project management:** Investment selection, resource allocation, and scheduling.
- **Product design:** Optimizing performance, cost, and reliability.
- **Supply chain management:** Network design, inventory management, and production planning.
- **Energy management:** Energy consumption optimization, cost reduction, and pollution mitigation.
- **Financial planning:** Budget allocation, risk management, and portfolio optimization.

In conclusion, MODM offers a powerful tool for addressing complex real-world problems. By systematically considering multiple objectives, decision-makers can make more informed and robust choices, leading to improved outcomes (Asgarpour, 2023).

1.6 Key Terms in MODM

In the realm of multi-objective decision-making (MODM), a wide array of terms and concepts is employed to describe various ideas. Below are some of the most crucial terms and their brief explanations:

1.6.1 Fundamental Concepts (Mehregan, 2013)

- **Objective:** Anything that a decision-maker seeks to achieve. In MODM, there are typically multiple objectives that may conflict with each other.
- **Constraint:** Limitations imposed on acceptable decisions.
- **Decision Variable:** Variables that can be altered by the decision-maker to attain desired objectives.
- **Decision space:** The set of all possible decisions.
- **Objective Space:** The set of all possible combinations of objective values.

1.6.2 Different Methods (Mehregan, 2013)

- **Weighted Sum Method:** Assigns weights to each objective and creates a single-objective function that is a linear combination of all objectives.

- **Goal Programming:** Assigns a target value to each objective and seeks to minimize the deviation between the actual and target values.
- **Ranking Methods:** Ranks objectives based on their importance and then uses methods to find efficient solutions based on this ranking.
- **Epsilon-Constraint Method:** Selects one objective as the primary objective and treats the rest as constraints.
- **Multi-criteria Decision Analysis (MCDA):** A set of methods for selecting the best option from among several options based on multiple criteria.
- **Multi-objective Optimization:** Aims to find a complete set of efficient solutions (Pareto optimal solutions).

1.6.3 Related Concepts (Mehregan, 2013)

- **Pareto Optimal Solution:** A solution where no other solution exists that can improve at least one objective without worsening another.
- **Pareto Front:** The set of all Pareto optimal solutions.
- **Trade-Off:** Improving one objective at the expense of another.
- **Weight:** A number representing the relative importance of each objective.
- **Goal:** The desired value for each objective.

1.6.4 More Specialized Terms (Mehregan, 2013)

- **Membership Function:** In fuzzy methods, a membership function is defined for each objective to indicate the degree to which a solution belongs to the fuzzy set of that objective.
- **Decision Matrix:** A table where options are listed in rows and criteria in columns, and each cell shows the value of that criterion for that option.
- **Euclidean Distance:** A measure of the distance between two points in the objective space.
- **MODM Software:** Software such as MATLAB, GAMS, LINGO, and specialized MODM software are used to solve multi-objective decision problems.
- **Decision Variable:** That which a decision-maker seeks to determine the optimal value of and whose value is determined by solving a model with multiple objectives and constraints.
- **Objective:** Objectives are created based on the decision-maker's perspective and using decision variables. In general, objectives represent the desires and needs of the decision-maker.
- **Goal (or Target):** Objectives do not inherently have specific levels. When levels are also considered for objectives, it is referred to as a goal or target.
- **Decision Alternatives:** Various solutions obtained by solving a model that the decision-maker can choose from are called decision alternatives.

- **Solution:** The set of values assigned to the decision variables. These solutions may or may not satisfy the objectives and constraints.
- **Feasible Solution:** A set of values that satisfy all the constraints of the problem is called a feasible solution.
- **Optimal Solution:** A feasible solution that optimizes all the objectives of the problem is called an optimal, ideal, or superior solution.
- **Dominant Solution:** A feasible solution A is said to dominate another feasible solution B if it is at least as good as B with respect to all objectives and strictly better than B with respect to at least one objective.
- **Preferred Solution:** A non-dominated solution that is selected by the decision-maker as the final solution based on other selection criteria.
- **Satisfactory Solution:** A solution that is chosen by the decision-maker and, in other words, achieves the decision-maker's desired levels for each of the objectives.

1.7 Evolution of Multi-Objective Decision-Making

The most basic allocation model is the classical allocation model, in which allocation is made based on a single objective, and its variables are binary (0 or 1). In this model, each object can be assigned to a task, and each task has only one requester. Therefore, two types of constraints exist in this model: the constraint of assigning each object to a task and the constraint of assigning each task to one individual.

In practice, decision-makers consider multiple, and sometimes conflicting, criteria rather than a single one. Accordingly, models have been designed that are commonly known as multi-criteria models. These models are broadly divided into two main categories: multi-objective models and multi-attribute models. Multi-objective models are used for design purposes, while multi-attribute models are employed for prioritizing, classifying, and generally ranking options (Hwang and Yoon, 1981).

Mathematical models of multi-criteria decision-making in their general form are as follows:

$$Max(or\ \ Min)\ \ Z = \left(Z_1, Z_2, Z_3, \ldots, Z_p\right)$$

where:

$$\begin{aligned} Z_1 &= Z_1\left(x_j\right) \\ Z_2 &= Z_2\left(x_j\right) \\ &\vdots \\ Z_p &= Z_p\left(x_j\right) \end{aligned}$$

subject to:

$$g_i(x_j) \leq b_i, \quad i = 1, 2, \ldots, m,$$
$$x_j \geq 0, \quad j = 1, 2, \ldots, n.$$

In this formulation:

- Z represents the vector of objective functions Z_1, Z_2, Z_3, …, Z_p which are functions of the decision variablesx_j.
- The objective is to either maximize or minimize these objective functions simultaneously.
- The constraints $g_i(x_j) \leq b_i$ impose limits on the decision variables, where b_i are the bounds and $g_i(x_j)$ are functions of the decision variables.
- $x_j \geq 0$ represents the non-negativity constraints on the decision variables.

This formulation captures the essence of multi-objective decision-making where trade-offs are made between conflicting objectives under a given set of constraints.

This model consists of p objective functions, n decision variables, and m constraints. By solving this model, the decision variables x_j are obtained, which satisfy all m constraints while optimizing the n objective functions.

Given that in many multi-objective decision-making problems, the objectives are often in conflict with each other, it is generally not possible to find an optimal solution that simultaneously optimizes all objectives. Therefore, the decision-maker typically selects a satisfactory solution.

In these models, the measurement scale for each objective may differ from that of other objectives, making it impossible to aggregate them. As a result, the objective function vector generally cannot be optimized simultaneously (Hamies and Hall, 1974).

To find the optimal solution, it is essential to have information about the available preferences. Without such information, the objectives may be conflicting and incomparable, making it impossible to achieve an optimal solution, as all feasible solutions cannot be compared. In this case, a complete ranking of the solutions can only be achieved by incorporating value judgments into the decision-making process. In many cases, a set of non-dominated solutions is sought. The set of non-dominated solutions related to the set of feasible solutions, denoted by X, is represented by S and is defined as follows:

$$S = \{x : x \in X\}, x' \in X \text{ such that } Z_q(x') > Z_q(x)$$
$$k \neq q \forall Z_k(x') \geq Z_k(x) \quad q \in \{1, 2, 3, \ldots, k\}.$$

A classification and comparison of multi-objective optimization methods in various resources such as Cohon and Marks (1975), Duckstein and Sudarovszky

(1994), and Fuller and Carlsson (1996) is reviewed. Cohon and Marks have proposed three criteria for evaluating multi-objective optimization techniques, namely, computational efficiency, explicit trade-offs between objectives, and the amount of information generated for decision-making. Based on these criteria, they categorized multi-objective optimization problems (with a single decision-maker) into three groups: (a) methods for generating non-dominated sets, (b) methods with a priori articulation of preference, and (c) methods with progressive preference articulation.

Among the methods in group A, weight assignment to objective functions, the min-max method, and the Philip and Zeleny method can be mentioned. In the latter two methods, there is no need to convert the multi-objective problem into a single-objective one; instead, the vector of objectives can be directly addressed to obtain non-dominated solutions. Both methods are exclusively applicable to linear problems. Among the methods in group B, goal programming methods, first introduced by Dantzig and Thapa, are now widely used (Charnes and Cooper, 1961).

The utility function assessment method, proposed by Keeney and Raiffa, also falls into group B (Keeney and Raiffa, 1993). A method called the trade-off analysis is also included in this group. Among the methods in group C, compromise programming can be mentioned, in which an ideal solution is specified. The solution that has the shortest distance to the ideal point is selected as the optimal solution (Zeleny, 1973). The aforementioned methods are suitable for solving continuous multi-objective problems.

Golany (1988) introduced an interactive multi-objective linear programming procedure to generate a set of alternative efficient points for decision-making units. Lozano and Villa (2009) proposed two distinct multi-objective data envelopment analysis (DEA) targeting approaches. The first method employs an interactive approach, while the second utilizes a lexicographic multi-objective approach. Malekmohammadi et al. (2009) applied centralized resource allocation in interval data envelopment analysis for performance evaluation and targeting of commercial banks in Malaysia. Lozano and Gutiérrez (2011) presented a multi-objective data envelopment analysis approach for targeting in the airline industry. Furthermore, the technical efficiency of each airline was evaluated using a composite-based efficiency measure. Kazemi Matin and Azizi (2011) introduced a new aggregation approach to set a nonnegative target related to the negative components for each observed unit. Triantis and Liu (2011) presented a data envelopment analysis-based approach for performance measurement and targeting in two real manufacturing organizations.

Some researchers have employed data envelopment analysis (DEA) in production planning. Du and Liang (2009), based on DEA and an extended Cournot model, determined how each decision-making unit selects its most preferred output production program when the market demand for one or more outputs is expected to change in the next production period. Du et al. (2010) investigated production planning in large organizations with centralized decision-making environments, assuming (1) the optimization of the average or total production performance of the organization and (2) the simultaneous maximization of total outputs and minimization of total inputs.

Building upon the work of Du et al. (2010), Hosseinzadeh Lotfi and Moghtaderi (2010) modeled how a central decision-maker determines new production plans for all units.

Amirteimoori and Kordrostami (2011) criticized the models presented in Du et al. (2010) and employed data envelopment analysis (DEA) to examine the production planning problem in a centralized decision-making environment where changes in the demand for products in the next production period can be forecasted. Amirteimoori and Kordrostami (2012) extended their previous studies on production planning based on DEA in centralized decision-making environments. The approach presented in their paper considers the size of operational units and determines the production level for each unit, taking into account the capabilities of each unit. Yuan (2012) proposed a two-phase fuzzy optimization approach for production planning that provides a novel perspective on analyzing multiproduct, multi-period problems. Their proposed approach aims to allocate the appropriate variety and quantity of products in each production period.

Many organizations presently have separate processes for long-term strategic planning and annual and short-term budgeting. Each organization has at least four types of resources that must be allocated to achieve its objectives: financial resources, physical resources, human resources, and technological resources. The efficient allocation of resources to achieve optimal levels of production, distribution, and consumption is linked to investment and pricing beyond the economics of perfect competition (Amirteimoori and Kordrostami, 2011).

Thanassoulis and Dyson (1992) presented goal programming as a problem related to the solutions of various mathematical programs. Their models utilize a preference structure that allows each decision-making unit to assign different levels of importance to potential changes in input/output levels.

Athanassopoulos (1998) proposed a goal programming model based on the multiplicative data envelopment analysis (DEA) model for resource allocation and goal setting. Athanassopoulos et al. (1999), building on the work of Thanassoulis and Dyson regarding goal setting for a decision-making unit under different assumptions, presented a paper. They applied their approach to data on electricity generation facilities in the UK. Athanassopoulos developed a DEA-based goal programming model for centralized planning, incorporating the ideas of Thanassoulis and Dyson (1992).

Amirteimoori and Emrouznejad (2011) proposed a data envelopment analysis (DEA)-based approach to determine the maximum possible reduction in a specific input variable and the minimum possible reduction in a specific output variable without reducing the efficiency of any decision-making unit (DMU). Lin (2011) presented a new approach for allocating fixed resources or costs by improving Cook and Zhu's (2010) approach and setting fixed targets based on the fixed resources shared by different DMUs. Lozano and Gutiérrez (2011) presented a centralized DEA-based approach for goal setting and resource allocation in Spain's national transportation system. Li et al. (2013) proposed several multi-objective linear programming models for resource allocation that consider reducing inputs, increasing desirable outputs, and decreasing undesirable outputs. They assumed that the unit's ability to change is based on the relative scale of current production and that undesirable outputs can be freely disposed of.

For example, a multi-objective linear programming model can be employed for resource allocation and goal setting in the following multi-component systems, as illustrated in a case study (Amirteimoori & Soufi, 2019):

$$Min \sum_{j=1}^{n} \left[b_j^+ - b_j^-\right]$$

$$Min \sum_{z=1}^{Z} \sum_{j=1}^{n} \left[a_{zj}^+ - a_{zj}^-\right]$$

$$Min \sum_{t=1}^{T} \sum_{j=1}^{n} \left[a_{tj}^+ - a_{tj}^-\right]$$

$$Min \sum_{r=1}^{s} \sum_{j=1}^{n} \left[a_{rj}^+ - a_{rj}^-\right]$$

$$Min \sum_{k=1}^{K} \sum_{j=1}^{n} \left[a_{kj}^+ - a_{kj}^-\right]$$

$$Min \sum_{d=1}^{D} \sum_{j=1}^{n} \left[a_{dj}^+ - a_{dj}^-\right]$$

$$Min \sum_{p=1}^{P} \sum_{j=1}^{n} \left[a_{pj}^+ - a_{pj}^-\right].$$

Moreover:

$$
\begin{aligned}
Min & \sum_{j=1}^{n}\left[b_j^+ - b_j^-\right] + \sum_{z=1}^{Z}\sum_{j=1}^{n}\left[a_{zj}^+ - a_{zj}^-\right] + \sum_{t=1}^{T}\sum_{j=1}^{n}\left[a_{tj}^+ - a_{tj}^-\right] \\
& + \sum_{r=1}^{s}\sum_{j=1}^{n}\left[a_{rj}^+ - a_{rj}^-\right] + \sum_{k=1}^{K}\sum_{j=1}^{n}\left[a_{kj}^+ - a_{kj}^-\right] + \sum_{d=1}^{D}\sum_{j=1}^{n}\left[a_{dj}^+ - a_{dj}^-\right] \\
& + \sum_{p=1}^{P}\sum_{j=1}^{n}\left[a_{pj}^+ - a_{pj}^-\right]
\end{aligned}
$$

$s.t.$

$$
\begin{aligned}
&\Big[\left(\sum\nolimits_{r=1}^{s} u_r^{(i)} y_{rj}^{(i)} + \sum\nolimits_{r=1}^{s} \overline{d}_{rj}^{i}\right)\left(\sum\nolimits_{i=1}^{b}\sum\nolimits_{r=1}^{s} u_r^{(i)} y_{rj}^{(i)} + \sum\nolimits_{i=1}^{b}\sum\nolimits_{r=1}^{s} \overline{d}_{rj}^{i}\right) \\
&+ \left(\sum\nolimits_{i=1}^{b}\sum\nolimits_{k=1}^{K} u_k^{(si)} \beta_i\left(y_{kj}^{(si)}\right) + \sum\nolimits_{i=1}^{b}\sum\nolimits_{k=1}^{K} \overline{\rho}_{kj}\right) \\
&- \left(\sum\nolimits_{i=1}^{b}\sum\nolimits_{z=1}^{Z} \rho_z^{(si)} \gamma_i\left(\overline{x}_{zj}^{(si)}\right)\sum\nolimits_{i=1}^{b}\sum\nolimits_{z=1}^{Z} \overline{g}_{zj}\right)\Big] \\
&- E_j^{(a)}\Big[\left(\sum\nolimits_{i=1}^{b}\sum\nolimits_{d=1}^{D} v_d^{(i)} x_{dj}^{(i)} + \sum\nolimits_{i=1}^{b}\sum\nolimits_{d=1}^{D} \overline{f}_{dj}^{i}\right) \\
&+ \left(\sum\nolimits_{i=1}^{b}\sum\nolimits_{p=1}^{P} v_p^{(si)}\left(\alpha_i x_{pj}^{(si)}\right) + \sum\nolimits_{i=1}^{b}\sum\nolimits_{p=1}^{P} \overline{c}_{pj}\right) \\
&+ \left(\sum\nolimits_{i=1}^{b}\sum\nolimits_{t=1}^{T} w_t^{(i)} \overline{y}_{tj}^{(i)} + \sum\nolimits_{i=1}^{b}\sum\nolimits_{t=1}^{T} \overline{\theta}_{tj}^{i}\right)\Big] \geq 0,
\end{aligned}
$$

$$
\begin{aligned}
&\Big[\left(\sum\nolimits_{r=1}^{s} u_r^{(i)} y_{rj}^{(i)} + \sum\nolimits_{r=1}^{s} \overline{d}_{rj}^{i}\right)\left(\sum\nolimits_{i=1}^{b}\sum\nolimits_{r=1}^{s} u_r^{(i)} y_{rj}^{(i)} + \sum\nolimits_{i=1}^{b}\sum\nolimits_{r=1}^{s} \overline{d}_{rj}^{i}\right) \\
&+ \left(\sum\nolimits_{i=1}^{b}\sum\nolimits_{k=1}^{K} u_k^{(si)} \beta_i\left(y_{kj}^{(si)}\right) + \sum\nolimits_{i=1}^{b}\sum\nolimits_{k=1}^{K} \overline{\rho}_{kj}\right) \\
&- \left(\sum\nolimits_{i=1}^{b}\sum\nolimits_{z=1}^{Z} \rho_z^{(si)} \gamma_i\left(\overline{x}_{zj}^{(si)}\right)\sum\nolimits_{i=1}^{b}\sum\nolimits_{z=1}^{Z} \overline{g}_{zj}\right)\Big]
\end{aligned}
$$

$$
\begin{aligned}
&- \Big[\left(\sum\nolimits_{i=1}^{b}\sum\nolimits_{d=1}^{D} v_d^{(i)} x_{dj}^{(i)} + \sum\nolimits_{i=1}^{b}\sum\nolimits_{d=1}^{D} \overline{f}_{dj}^{i}\right) \\
&+ \left(\sum\nolimits_{i=1}^{b}\sum\nolimits_{p=1}^{P} v_p^{(si)}\left(\alpha_i x_{pj}^{(si)}\right) + \sum\nolimits_{i=1}^{b}\sum\nolimits_{p=1}^{P} \overline{c}_{pj}\right) \\
&+ \left(\sum\nolimits_{i=1}^{b}\sum\nolimits_{t=1}^{T} w_t^{(i)} \overline{y}_{tj}^{(i)} + \sum\nolimits_{i=1}^{b}\sum\nolimits_{t=1}^{T} \overline{\theta}_{tj}^{i}\right)\Big] \\
&= b_j^+ - b_j^-,
\end{aligned}
$$

$$
\begin{aligned}
&\overline{g}_{zj} - \Delta_j^z \rho_z^{(si)} G_z = a_{zj}^+ - a_{zj}^-, \\
&\overline{\theta}_{tj}^i - \Delta_j^t w_z^{(i)} \Theta_t^i = a_{tj}^+ - a_{tj}^-, \\
&\overline{d}_{rj}^i - \Delta_j^r u_r^{(i)} D_r^i = a_{rj}^+ - a_{rj}^-, \\
&\overline{p}_{kj} - \Delta_j^k u_k^{(si)} P_k = a_{kj}^+ - a_{kj}^-, \\
&\overline{f}_{dj}^i - \Delta_j^d v_d^{(i)} F_d^i = a_{dj}^+ - a_{dj}^-, \\
&\overline{c}_{pj} - \Delta_j^p v_p^{(si)} C_p = a_{pj}^+ - a_{pj}^-, \\
&\sum_{i=1}^{b} \alpha_i = 1, \\
&\sum_{i=1}^{b} \beta_i = 1, \\
&\sum_{i=1}^{b} \gamma_i = 1, \\
&u_r^{(i)}, u_k^{(si)}, \rho_z^{(si)}, v_d^{(i)}, v_p^{(si)}, w_z^{(i)} \geq \varepsilon, \\
&\overline{g}_{zj} \geq 0 \ when \ G_z \geq 0 \& \ \overline{g}_{zj} \leq 0 \ when \ G_z \leq 0, \\
&\overline{\theta}_{tj}^i \geq 0 \ when \ \Theta_t^i \geq 0 \& \overline{\theta}_{tj}^i \leq 0 \ when \ \Theta_t^i \leq 0, \\
&\overline{d}_{rj}^i \geq 0 \ when \ D_r^i \geq 0 \& \overline{d}_{rj}^i \leq 0 \ when \ D_r^i \leq 0, \\
&\overline{p}_{kj} \geq 0 \ when \ P_k \geq 0 \& \overline{p}_{kj} \leq 0 \ when \ P_k \leq 0, \\
&\overline{f}_{dj}^i \geq 0 \ when \ F_d^i \geq 0 \& \overline{f}_{dj}^i \leq 0 \ when \ F_d^i \leq 0, \\
&\overline{c}_{pj} \geq 0 \ when \ C_p \geq 0 \& \ \overline{c}_{pj} \leq 0 \ when \ C_p \leq 0.
\end{aligned}
$$

In addition, these coefficients, defined by the decision-maker and representing the relative importance of objectives, can be incorporated into the model's objective function. It should be noted that, according to the proposed program, changes in the desired and undesired outputs produced by each unit, as well as the total consumption of inputs, will be allocated proportionally among the different units based on the relative size of each unit, in a manner that does not reduce the efficiency of any unit. The new production possibility set will serve as a reliable reference for future decision-making.

1.8 Methods for Solving MODM Problems

In multi-objective decision-making problems, due to the conflicting nature of objectives, a single optimal solution typically does not exist. Instead, a set of Pareto optimal solutions is obtained, where improving one objective necessarily leads to a deterioration in at least one other objective.

A variety of methods exist for addressing multi-objective decision-making problems. In subsequent sections, several of these approaches will be discussed, including:

- Constraint-based methods.
- Weighted sum method.
- Absolute priorities method.
- Global criterion method.
- STEM method.
- Goal programming.

Before presenting the problem using the method of converting objectives into constraints, a single-objective problem is presented to familiarize the reader with the general form of such problems:

Example 1.1

Consider the following multi-objective programming model

$$
\begin{aligned}
&Max\ f_1 = x_1 + x_2\\
&Max\ f_2 = x_2\\
&s.t.\\
&x_1 \leq 3,\\
&x_2 \leq 3,\\
&x_1, x_2 \geq 0.
\end{aligned}
$$

Solution: The above model is solved as a single-objective model for each objective individually, and the optimal solution for each objective is obtained. Since the solution for each objective is optimal only for that specific objective, it is considered a non-dominated solution. Finally, the optimal solution is computed from among the non-dominated solutions, as shown in Table 1.1.

Table 1.1 Solutions for Example 1.1

	Solution	Type
First objective	$f_1 \rightarrow$ B(3,3) $\rightarrow f_1 = 6$	Non-dominated solution
Second objective	$f_2 \rightarrow$ B(0,3) $\rightarrow f_2 = 3$	Non-dominated solution
First objective	$f_1 \rightarrow$ B(0,3) $\rightarrow f_1 = 3$	Optimal solution
Second objective	$f_2 \rightarrow$ B(0,3) $\rightarrow f_2 = 3$	Optimal solution

1.8.1 Constraint Transformation

In this method, decision-makers have the flexibility to specify maximum or minimum values for each of their objectives. These objectives are then incorporated into the problem as constraints, transforming the multi-objective problem into a single-objective one. The resulting single-objective problem can be solved using standard linear programming techniques. It is crucial to note that when converting goals to constraints, a logical process should be followed. The maximum or minimum values for the goals should be determined rationally, considering the decision-maker's judgment and sensitivity analysis.

Example 1.2
Consider the following multi-objective decision-making problem

$$\begin{aligned}
&Max\ f_1 = 3x_1 + 6x_2\\
&Max\ f_2 = 9505x_1 + 5450x_2\\
&s.t.\\
&4x_1 + x_2 \leq 12,\\
&x_1 \quad\ \ \leq 7,\\
&x_1, \quad x_2 \geq 0.
\end{aligned}$$

If the first objective (f_1) is related to the level of employment and the second objective (f_2) is related to the production cost, and the producer desires to keep costs below 120,000, solve the problem using the method of converting the objective into a constraint.

Solution: Given that the maximum allowable cost for the company is 120,000, a constraint is added to the problem, setting the second objective to be less than or equal to 120,000. This creates a linear programming model that can be easily solved:

$$\begin{aligned}
&Max\ f_1 = 3x_1 + 6x_2\\
&Max\ f_2 = 9505x_1 + 5450x_2\\
&s.t.\\
&4x_1 \quad + \quad x_2 \leq 12,\\
&x_1 \qquad\qquad \leq 7,\\
&9505x_1 + 5450x_2 \leq 120000\\
&x_1, \qquad\quad x_2 \geq 0.
\end{aligned}$$

The optimal solution to the linear program reveals that the decision variables x_1 and x_2 should be assigned values of 1.25 and 7, respectively. This allocation maximizes the objective function at a value of:

$$x_1 = \frac{1}{25},\ x_2 = 7,\ f_1 = \frac{45}{75},\ f_2 = \frac{50031}{25}.$$

1.8.2 Weighted Sum Method

In many decision-making problems involving multiple objectives, the objectives are not weighted equally. Decision-makers assign different weights to each objective. By considering these weights, multiple objectives can be combined into a single objective. Using linear programming techniques, the values of decision variables and, consequently, the values of each objective can be obtained.

When employing the weighted sum method for multi-objective optimization, the following points must be considered:

1. Weights must be dimensionless. The sum of the weights of all objectives should equal 1:

$$\sum_{i=1}^{n} w_i = 1.$$

2. All objectives should be either maximized or minimized. If this condition is not met, it can be adjusted as follows: $Max\,Z = Max\,(-Z)$.
3. Coefficients of decision variables in each objective should also be normalized.

Example 1.3
An alloy manufacturer receives a customer order for the production of an alloy as shown in Table 1.2

In the production of the desired alloy, the metallic composition must adhere to the following constraints:

- The percentage of metal A must be at least 17%.
- The percentage of metal B must not exceed 21%.

Table 1.2 Composition percentage and unit selling price

Alloy sample	% Metal A	% Metal B	% Metal C	% Metal D	% Impurities	Price per ton ($)
1	23	10	10	25	34	23
2	42	0	0	32	26	20
3	20	10	0	30	40	18
4	0	25	5	10	60	10
5	25	23	0	41	11	27
6	8	5	10	17	60	12

- The percentage of metal C must not exceed 8%.
- The percentage of metal D must be between 31% and 61%.

Given the compositional data in Table 1.2, the producer aims to optimize three objectives:

1. Minimize the total cost per ton of the alloy.
2. Maximize extraction from mine 3 due to its proximity to the factory.
3. Minimize extraction from mine 6 due to its greater distance.

Employing the weighted sum method, with weights of 0.21, 0.42, and 0.28 assigned to the respective objectives, determine the optimal values for each objective and decision variable.

Solution: Initially, decision variables are defined, representing the quantity extracted from each of the six mines, denoted as x1 to x6. Subsequently, the initial model is formulated:

$$\begin{aligned}
&Max\ Z_1 = 23x_1 + 20x_2 + 18x_3 + 10x_4 + 27x_5 + 12x_6\\
&Max\ Z_2 = x_3\\
&Max\ Z_3 = x_6\\
&s.t.\\
&0.23x_1 + 0.42x_2 + 0.20x_3 + 0.00x_4 + 0.25x_5 + 0.08x_6 \geq 0.17\\
&0.10x_1 + 0.00x_2 + 0.10x_3 + 0.25x_4 + 0.23x_5 + 0.05x_6 \leq 0.21\\
&0.10x_1 + 0.00x_2 + 0.00x_3 + 0.05x_4 + 0.00x_5 + 0.10x_6 \leq 0.08\\
&0.25x_1 + 0.32x_2 + 0.30x_3 + 0.10x_4 + 0.41x_5 + 0.17x_6 \geq 0.31\\
&0.25x_1 + 0.32x_2 + 0.30x_3 + 0.10x_4 + 0.41x_5 + 0.17x_6 \leq 0.6\\
&0.34x_1 + 0.26x_2 + 0.40x_3 + 0.60x_4 + 0.11x_5 + 0.60x_6 = 1\\
&x_1, x_2, x_3, x_4, x_5, x_6 \geq 0.
\end{aligned}$$

Subsequently, the coefficients of the decision variables in the first objective and the weights of the objectives are normalized as follows:

$$\begin{array}{lll}
w_1 = 0.21 & w_2 = 0.18 & w_3 = 0.16\\
w_4 = 0.01 & w_5 = 0.24 & w_6 = 0.11\\
w_{z1} = 0.75 & w_{z2} = 0.15 & w_{z3} = 0.10
\end{array}$$

Furthermore, the second objective should also be minimized so that all three objectives are aligned:

$$Min\ -Z_2 = -x_3.$$

The multi-objective problem is transformed into a single-objective problem by aggregating the normalized objectives using a weighted sum:

$$
\begin{aligned}
&Min\ Z = 0.75(0.21x_1 + 0.18x_2 + 0.16x_3 + 0.01x_4 + 0.24x_5 + 0.11x_6) + 0.15(-x_3) + 0.10x_6\\
&s.t.\\
&0.23x_1 + 0.42x_2 + 0.20x_3 + 0.00x_4 + 0.25x_5 + 0.08x_6 \geq 0.17\\
&0.10x_1 + 0.00x_2 + 0.10x_3 + 0.25x_4 + 0.23x_5 + 0.05x_6 \leq 0.21\\
&0.10x_1 + 0.00x_2 + 0.00x_3 + 0.05x_4 + 0.00x_5 + 0.10x_6 \leq 0.08\\
&0.25x_1 + 0.32x_2 + 0.30x_3 + 0.10x_4 + 0.41x_5 + 0.17x_6 \geq 0.31\\
&0.25x_1 + 0.32x_2 + 0.30x_3 + 0.10x_4 + 0.41x_5 + 0.17x_6 \leq 0.6\\
&0.34x_1 + 0.26x_2 + 0.40x_3 + 0.60x_4 + 0.11x_5 + 0.60x_6 = 1\\
&x_1, x_2, x_3, x_4, x_5, x_6 \geq 0.
\end{aligned}
$$

Solving the model yields the following decision variable values:

$$x_1 = 0, x_2 = 0, x_3 = 1.719, x_4 = 0.0603, x_5 = 0, x_6 = 0.4603.$$

The corresponding objective function values are:

$$Z_1 = 37.0686,\ \ Z_2 = 1.714,\ \ Z_3 = 0.4603.$$

1.8.3 Absolute Priorities Method

Another approach to solving multi-objective decision-making problems is the method of absolute priority. In this method, the decision-maker must be able to prioritize their objectives and enforce this prioritization strictly. To this end, the objective with the highest priority is selected and solved, given the problem's constraints. The resulting solution is then incorporated as a constraint in the problem, and the objective with the next highest priority is considered. This process continues until all objectives have been evaluated.

Example 1.4
A manufacturing organization produces two types of products. The processing time for each product is 15 and 23 minutes, respectively, and the labor required for each unit of product is 6 and 4 people, respectively. Similarly, the profit from producing each product is 65 and 58 units, respectively. The organization will start production only if the minimum profit from producing products is equal to 15,050 monetary units. Also, the maximum demand for the first product is 250 and for the second product is 150. The organization is initially looking to reduce the processing time and then, as a second priority, to reduce the required labor. Given this problem and using the absolute priority method, determine the value of each decision variable and objective.

Solution: Initially, the decision variables are defined as the production quantities of each product and are denoted by x_1 and x_2. Subsequently, a multi-objective optimization problem is formulated, as detailed below:

$$\begin{aligned}
&Min\ f_1 = 15x_1 + 23x_2\\
&Min\ f_2 = 6x_1 + 4x_2\\
&s.t.\\
&65x_1 + 58x_2 \leq 15050,\\
&x_1 \qquad\quad \leq 250,\\
&\qquad\quad x_2 \leq 150,\\
&x_1, \quad x_2 \geq 0.
\end{aligned}$$

Given that minimizing processing time is the primary objective, the linear programming model with the first objective function and the problem's constraints is solved initially, and the value of the objective function is determined:

$$\begin{aligned}
&Min\ f_1 = 15x_1 + 23x_2\\
&s.t.\\
&65x_1 + 58x_2 \leq 15050,\\
&x_1 \qquad\quad \leq 250,\\
&\qquad\quad x_2 \leq 150,\\
&x_1, \quad x_2 \geq 0.
\end{aligned}$$

The solution of this model yields the following values for the variables and the objective function:

$$x_1 = 231.538, \qquad x_2 = 0, \qquad f_1 = 3473.077.$$

The optimal value of the first objective function is then added as a constraint, and the linear programming model is solved with the second objective function to obtain the solution:

$$
\begin{aligned}
& Min\ f_2 = 6x_1 + 4x_2 \\
& s.t. \\
& 15x_1 + 23x_2 \leq 3473.077, \\
& 65x_1 + 58x_2 \leq 15050, \\
& x_1 \qquad\quad \leq 250, \\
& \qquad\quad x_2 \leq 150, \\
& x_1, \qquad x_2 \geq 0.
\end{aligned}
$$

By solving the above model, the problem's solutions are calculated as follows:

$$x_1 = 231.538, \quad x_2 = 0, \quad f_1 = 3473.077, \quad f_2 = 1389.231.$$

1.8.4 Global Criterion Method

The global criterion method is a technique used in multi-objective optimization to transform a problem with multiple conflicting objectives into a single-objective optimization problem.

In this method, multiple objectives to be optimized are clearly defined. *Here, we're establishing that there are several goals that we want to make as good as possible.*

Then, a weight is assigned to each objective to reflect its relative importance. *This step involves giving a numerical value to each objective, indicating how much it matters compared to the others.*

Subsequently, the individual objective functions are combined into a single global criterion function using a weighted sum or other suitable aggregation method. *Here, we're taking all the different objective functions and combining them into one overall function. This is done by multiplying each objective function by its corresponding weight and then adding them all up.*

Finally, the resulting single-objective optimization problem is solved to find the optimal solution. *This step involves using mathematical techniques to find the best possible solution, considering the combined objective function.*

As previously discussed, in multi-objective problems, objectives can conflict with one another, and optimizing one objective may lead to a suboptimal solution for another. Given this challenge, the comprehensive criterion method aims to minimize the relative deviations of objectives from their optimal values. After solving the problem for each objective individually, the following objective function is constructed, with the goal of minimizing it:

$$Min\ Z = \sum_{i=1}^{k} \left(\frac{f_i^*}{f_i^*} \right)^p$$

where f_i^* is the optimal value of the (i)th objective function, considering other objectives independently. The parameter (p) is an exponent applied to the sum of relative deviations from the optimal values, and it can take values of 1, 2, 3, … .

If $p = 1$, the model is linear; otherwise, it is nonlinear.

Example 1.5

Consider the following multi-objective problem designed for airline customer service planning. The first objective is to increase flight safety, and the second is to increase the speed of service delivery. Using the comprehensive criterion method, determine the values of the decision variables and objectives:

$$\begin{aligned}
&Max\ f_1 = 1245x_1\\
&Max\ f_2 = 0.68x_1 + 0.78x_2\\
&s.t.\\
&10x_1 + 31x_2 \leq 450,\\
&x_1 \qquad\quad \geq 25,\\
&2x_1 + x_2 \quad \leq 64,\\
&x_1, \qquad x_2 \geq 0.
\end{aligned}$$

Solution: In the first step, the problem is individually optimized for each objective function. This process yields the optimal value for each objective, which can be expressed as follows:

$$\begin{array}{lll}
x_1 = 32, & x_2 = 0, & f_1^* = 39840\\
x_1 = 29.5, & x_2 = 0, & f_1^* = 23.96
\end{array}$$

Subsequently, the relative deviation function from the objectives, considering P = 1, is expressed as follows:

$$\begin{aligned}
&Min\ Z = \left[\frac{39840 - 12451x_1}{39840}\right] + \left[\frac{23.96 - 0.68x_1 - 0.78x_2}{23.96}\right]\\
&s.t.\\
&10x_1 + 31x_2 \leq 450,\\
&x_1 \geq 25,\\
&2x_1 + x_2 \leq 64,\\
&x_1, x_2 \geq 0.
\end{aligned}$$

By solving this model, the values of the decision variables are obtained. Using these values, the value of each objective is calculated, as follows:

$$x_1 = 29.6, \qquad x_2 = 5,$$
$$f_1 = 36727.5, \qquad f_2 = 23.96$$

Note: In the global criterion method, a weight can be assigned to each objective of the problem. In this case, the objective function is formulated as follows:

$$Min\ Z = \sum_{i=1}^{k} \left(\frac{f_i^*}{f_i^*}\right)^p.$$

The sum of the weights assigned to the objectives in the aforementioned relation is equal to 1.

1.8.5 STEM Method

Various techniques and approaches have been presented to solve multi-objective decision-making problems under different assumptions. The STEM technique is also one of the most widely used techniques for dealing with this group of problems, especially for solving multi-objective linear programming (MOLP) models.

One of the most significant and well-known techniques within the realm of interactive multi-objective decision-making is the STEM method, originally introduced by Benayoun and his colleagues in 1971. Step method (STEM) is one of the well-known multi-objective optimization techniques. STEM has proven to be effective in extracting a decision-maker (DM)'s preference information for a satisfactory compromise. However, it has been criticized for not considering the differing degrees of satisfaction associated with an objective function value and for not providing flexible options in the process of preference information extraction. This method is renowned as the first interactive approach for solving multi-objective optimization problems. Since its inception, STEM has been widely applied to address a vast array of decision-making problems across various domains.

To solve multi-objective problems using this method, the following steps are involved:

$$\begin{aligned} &Max\ C_1x \\ &Max\ C_2x \\ &\vdots \\ &s.t.\ S^m \begin{cases} Ax \geq b \\ x_j \geq 0 \end{cases} \end{aligned}.$$

Step 1: Solve each objective of the problem separately and calculate the optimal value for each:

Table 1.3 Productivity matrix

	f_1	....	f_j	....	f_k
f_1	f_1^*	....	Z_{ij}	....	Z_{1k}
•	•	....	•	....	•
•	•		•		•
f_j	Z_{j1}	....	f_j^*	....	Z_{jk}
•	•	....	•	....	•
•	•		•		•
f_k	Z_{k1}	....	Z_{kj}	....	f_k^*

$$\begin{aligned} & Max\, f_j = c_j x \quad j = 1, 2, \ldots, n \\ & s.t. S^m \end{aligned}$$

In this case, x_j^* denotes the optimal value of x_j for the jth objective, generating the objective function f_j^*.

Step 2: Using the optimal values of the decision variables for each objective, a payoff matrix is constructed, as shown in Table 1.3. The diagonal elements of this matrix represent the optimal values of the corresponding objective functions. Specifically, the diagonal element Z_{ji} represents the optimal value of the jth objective function, $f_j(x)$. The off-diagonal element Z_{ji} represents the value of the jth objective function evaluated at the optimal solution of the ith objective function, i.e., $f_j(x)$.

Step 3: In this step, the following *Min* problem is solved to determine the closest possible distance to f_j^*:

$$\begin{aligned} & Min\ \gamma \\ & s.t. \left(f_j^* - f_j(x)\right).\beta_j \le \gamma, \quad j = 1, 2, \ldots, k, \\ & S^m \\ & \gamma \ge 0. \end{aligned}$$

In the linear programming model above, the value of β_j is obtained from the following relationship:

$$\beta_j = \frac{\alpha_j}{\sum_{j=1}^{k} \alpha_j}$$

$$\alpha_j = \frac{f_j^* - f_j^{\min}}{f_j^*} \left[\frac{1}{\sqrt{\sum_{i=1}^{n} C_{ij}^2}} \right] \quad if\ \ f_j^* > 0.$$

In the above relationship, $f_j^{\min}$ represents the minimum value in the *j*th row of the payoff matrix, and C_{ij} denotes the profit coefficients of the jth objective function:

$$\alpha_j = \frac{f_j^* - f_j^{\min}}{f_j^*}\left[\frac{1}{\sqrt{\sum_{i=1}^{n} C_{ij}^2}}\right] \qquad \textit{if} \;\; f_j^* \leq 0.$$

Example 1.6
Consider the following multi-objective problem and the optimal solutions for each objective

$$\begin{aligned}
&Max\ f_1 = x_1 - 3x_2 \quad f_1^* = 4 \quad x_1 = 4 \quad x_2 = 0\\
&Max\ f_2 = x_1 + 2x_2 \quad f_2^* = 10 \quad x_1 = 0 \quad x_2 = 5\\
&Max\ f_3 = 2x_1 + 3x_2 \quad f_3^* = 16 \quad x_1 = 2 \quad x_2 = 4\\
&s.t.\\
&x_1 + 3x_2 \leq 10,\\
&2x_1 + x_2 \leq 8,\\
&x_1, x_2 \geq 0
\end{aligned}$$

Solution: Formulate the problem using the STEM method:

	f_1	f_2	f_3
f_1	4	−15	−10
f_2	−4	10	6
f_3	8	15	16

$$\alpha_1 = \frac{4+15}{4}\left[\frac{1}{\sqrt{1+9}}\right] = 1.502$$

$$\alpha_2 = \frac{10+4}{10}\left[\frac{1}{\sqrt{1+4}}\right] = 0.626$$

$$\alpha_3 = \frac{16-8}{16}\left[\frac{1}{\sqrt{4+9}}\right] = 0.139$$

$$\sum \alpha_1 = 1.502 + 0.626 + 0.139$$

$$\beta_1 = 0.662$$

$$\beta_2 = 0.276$$

$$\beta_3 = 0.062$$

$$
\begin{aligned}
&Min\ \gamma \\
&s.t. \\
&\gamma \geq (4 - x_1 + 3x_2)0.662 \\
&\gamma \geq (10 + x_1 - 2x_2)0.276 \\
&\gamma \geq (16 - 2x_1 - 3x_2)0.062 \\
&x_1 + 3x_2 \leq 10 \\
&2x_1 + x_2 \leq 8 \\
&x_1, x_2 \geq 0
\end{aligned}
$$

After solving the above model, the problem's solutions are calculated as follows:

$$x_1 = 0 \quad \text{and} \quad x_2 = 0.044.$$

1.8.6 Goal Programming

Goal programming, a powerful technique for solving multi-objective decision-making problems, has its roots in the mid-twentieth century. It emerged as a response to the need to address complex, real-world problems where finding a single optimal solution that simultaneously satisfies all objectives was often infeasible. Goal programming, as a powerful tool for solving multi-objective decision-making problems, has evolved over several decades and found widespread application in various fields. Given the increasing complexity of real-world decision-making problems, goal programming is expected to remain a valuable tool in the future.

1.8.6.1 Development of Goal Programming Models

- **1950s:** The initial ideas related to goal programming were formed within the early research in linear programming and mathematical optimization. Researchers sought a method to model and solve problems with multiple, often conflicting, objectives.
- **1960s:** The concept of goal programming was formally introduced through the publications of Charnes and his colleagues. They developed this approach to address decision-making problems with multiple goals and constraints.
- **1970s:** Goal programming expanded rapidly and found applications in various fields such as management, economics, engineering, and social sciences. Researchers proposed new and improved methods for goal programming.
- **1980s and Beyond:** With advancements in computers and optimization software, goal programming became a highly efficient tool for solving complex problems.

Additionally, the integration of goal programming with other decision-making techniques, such as fuzzy programming, expanded its applications.
Charnes et al. introduced the concept of deviation, which represents the difference between a goal's actual value and its desired value. To determine the relative importance of different goals, a weight is assigned to each goal.

Various methods for solving goal programming models were developed, including linear, nonlinear, and fuzzy approaches. Goal programming was applied in various fields such as production planning, project management, product design, finance, and investment.

In recent years, goal programming continues to evolve, and researchers are exploring ways to improve its methods and expand its applications. Some of these developments include:

- **Integration with other techniques:** Combining goal programming with artificial intelligence, machine learning, and data analysis techniques.
- **Applications in large-scale and complex problems:** Developing efficient algorithms for solving goal programming problems with a large number of variables and constraints.
- **Applications in problems with uncertainty:** Using fuzzy and probabilistic goal programming to model problems with incomplete or vague information.

1.8.6.2 Key Concepts in Goal Programming

1. **Goals:** These are the specific objectives that a decision-maker aims to achieve. Goals can be quantitative (e.g., maximizing profit) or qualitative (e.g., improving product quality).
2. **Deviations:** Deviations represent the difference between a goal's actual value and its desired value. Deviations can be positive (when the actual value exceeds the desired value) or negative (when the actual value falls short of the desired value).
3. **Weights:** Each goal is assigned a weight to indicate its relative importance. Higher weights signify more critical goals.
4. **Objective Function:** This is a mathematical function that minimizes the weighted sum of deviations. The aim is to minimize the overall deviation from the desired goals.
5. **Mathematical Model:** A mathematical representation of the problem, including decision variables, constraints, and the objective function.
6. **Optimal Solution:** The solution that minimizes the objective function while satisfying all constraints.

1.8.6.3 General Steps in Goal Programming

1. Identify Goals: Clearly define the objectives of the problem.
2. Assign Weights: Assign weights to each goal to reflect their relative importance.

3. Formulate the Model: Develop a mathematical model of the problem, including decision variables, constraints, and the objective function.
4. Solve the Model: Use optimization software to solve the model and obtain the optimal solution.
5. Interpret Results: Analyze the results and select the best solution based on the decision-maker's preferences.

1.8.6.4 Mathematical Model of Goal Programming

$$
\begin{aligned}
& Min\ Z = \sum\nolimits_{k=1}^{q} \sum\nolimits_{i=1}^{m} p_k^{\left(d_i^- + d_i^+\right)} \\
& s.t. \\
& \sum\nolimits_{j=1}^{m} C_{ij} - d_i^+ + d_i^- = b_i, \quad i = 1, 2, \ldots., m, \\
& \sum\nolimits_{j=1}^{s} a_{rj} x_j \le b_j, \quad j = 1, 2, \ldots, s, \\
& x_j, d_i^+, d_i^- \ge 0.
\end{aligned}
$$

In the above equation, "*m*" signifies the number of objectives, "*s*" represents the number of system constraints, and p_k denotes the priority associated with the "*k*"th objective.

1.8.6.5 Types of Goal Programming Models

- ***Weighted Goal Programming:*** In this approach, each objective is assigned a weight that reflects its relative importance. Subsequently, the weighted sum of deviations from the goals is minimized.
- ***Preemptive Goal Programming:*** In this approach, objectives are ranked based on their importance, and higher-priority goals are given precedence. In other words, the model first attempts to achieve the highest priority goals as much as possible and then proceeds to lower-priority goals.
- ***Lexicographic Goal Programming***: This approach combines the previous two methods. It first satisfies the highest priority goals absolutely and then optimizes the lower-priority goals using weights.
- ***Goal Programming with Hard Constraints:*** In this type, constraints are strict and cannot be violated.
- ***Goal Programming with Soft Constraints:*** In this type, constraints can be violated to some extent, but the cost of violation is considered.
- ***Goal Programming with a Linear Objective Function:*** In this type, the objective function is a linear function of deviations from the goals.
- ***Goal Programming with a Nonlinear Objective Function:*** In this type, the objective function can be a nonlinear function of deviations from the goals.

A variety of software tools are available to solve goal programming problems. General-purpose optimization software such as LINGO, CPLEX, and Excel Solver can be employed, while specialized goal programming software offers tailored algorithms and features for this specific problem type.

1.9 Case Study: This Goal Programming Model Has Been Designed for Production Planning of Precast Concrete Walls in a Concrete Company

The mathematical model developed for this research is a goal programming model. This model comprises 276 decision variables, two objective functions aiming to minimize costs and maximize sales, and six constraint sets. These constraints include capacity constraints for machines (ensuring that no machine produces more than its capacity), balancing constraints (linking production levels across lines, ensuring that the production quantity of each product at each station aligns with its required quantity for the subsequent production stage), demand constraints (inventory of a product in a given month equals the previous month's inventory plus production in that month minus the forecasted demand), overtime and regular working hour constraints (limiting working hours within the company's defined range), overtime personnel constraints (ensuring that the number of overtime workers does not exceed the total workforce), and inventory level constraints (limiting inventory levels to predefined maximums). To formulate capacity and balancing constraints, the production stages must be considered, and corresponding constraints must be defined for each station. Figure 1.1 illustrates the production stages.

1.9.1 Decision Variables

$f_1, f_2, f_3, f_4, f_5, f_6$: Excess costs incurred for products in October, November, December, January, February, and March, respectively

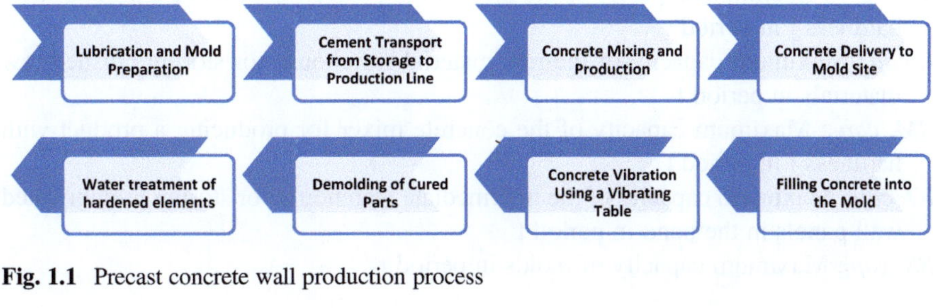

Fig. 1.1 Precast concrete wall production process

$h_1, h_2, h_3, h_4, h_5, h_6$: Shortfalls in sales targets for products in October, November, December, January, February, and March, respectively
PC_{1ijt}: Quantity of concrete produced in batch i with hardness level j in period t
PC_{2ijt}: Quantity of product produced in batch i with hardness level j in period t
IP_t: Amount of cement stored in the raw material warehouse in period t
RM_{jt}: Quantity of concrete produced with hardness level j in period t
RS_t: Quantity of concrete stored in period t
IO_t: Amount of clinker stored in the intermediate product warehouse in period t
CM_{ijt}: Number of walls produced by grinder i with type j in period t
ZL_t: Amount of additives used for product production in period t
IL_t: Number of walls stored in warehouse 1 in period t
IM_t: Number of walls stored in warehouse 2 in period t
IS_t: Number of walls stored in warehouse 3 in period t
S_{ijt}: Number of walls ready for sale from the final station with label i and type j in period t
N_t: Number of regular working hours for each worker in period t
L_t: Number of workers working overtime in period t
M_t: Number of overtime hours for each worker in period t
IS_{ijt}: Ending inventory of walls with label i and type j in period t.

1.9.2 Model Parameters

Model parameters include forecasted demand for various products based on product type and time period, maximum capacity of bands, and efficiency of different devices. Initial inventory of product i of type j in period t is another parameter. The determined targets for sales and costs, as well as the amount of regular wages and overtime pay, the cost of storing 1 square meter of product, the number of workers working overtime per month, the minimum number of regular working hours, the maximum and minimum number of overtime hours per worker, and the selling price and production cost of each square meter of product, are all considered model parameters.

$C.\ cap_{1ijt}$: Maximum capacity of the ith crusher to produce a product with hardness j in period t
$M.\ cap_{2ijt}$: Maximum capacity of the ith vibrating screen to produce a product with hardness j in period t
$I.\ cap_t$: Maximum capacity of the raw material warehouse for storing crushed raw materials in period t
$RM.\ cap_{jt}$: Maximum capacity of the concrete mixer for producing a product with hardness j in period t
$IO.\ cap_t$: Maximum capacity of the intermediate warehouse for storing prefabricated wall panels in the band in period t
$RS.\ cap_t$: Maximum capacity of molds in period t

$K.\ cap_t$: Maximum capacity of the ith kiln to produce a product with hardness j in period t

$CM.\ cap_{ijt}$: Maximum capacity of the ith vibrating screen for producing a product with hardness j in production period t

$IL.\ cap_t$: Maximum capacity of the first vibrating table for concrete in period t

$IM.\ cap_t$: Maximum capacity of the second vibrating table for concrete in period t

$IS.\ cap_t$: Maximum capacity of the third vibrating table for concrete in period t

$S.\ cap_{ijt}$: Maximum capacity of the ith sales station for product with label i and type j in period t

N_t min: Minimum regular hours that each worker works in period t

N_t max: Maximum regular hours that each worker works in period t

$L_t$max: Maximum number of workers working in period t

$M_t$max: Maximum overtime hours that each worker can work in period t

M_t min: Minimum overtime hours that each worker can work in period t

$LS_{ijt} - 1$: Inventory level of wall i with type j at the beginning of month t

D_{ijt}: Demand for prefabricated wall i with type j in period t

C_{ijt}: Cost of 1 square meter of prefabricated wall i with type j in period t

P_{ijt}: Selling price of 1 square meter of product i with type j in period t

C_{RL}: Cost of 1 hour of regular labor for each worker in period t

C_{RO}: Cost of 1 hour of overtime labor for each worker in period t

C_{IO}: Cost of storing 1 square meter of product in period t

g_t: Target for maximizing sales revenue

%X: Efficiency of the concrete production section

%Y: Efficiency of the concrete vibrating section

%A: Percentage of concrete

%(1-A): Percentage of additives.

1.9.3 Mathematical Model of the Problem

The investigation considers dual objectives: cost reduction and revenue enhancement. The constructed model expresses these goals through the subsequent equations.

In particular, the equation below accounts for four cost components: (1) production tier costs, (2) standard labor expenses, (3) overtime workforce expenditures, and (4) inventory storage costs, where parameter C_t denotes the aggregate cost target:

$$Z_1t = \sum_{i=1}^{m}\sum_{j=1}^{n} S_{ij}t^* C_{ij}t + \sum_{t=1}^{t} (L_t\max * N_t^* C_R l) + \sum_{t=1}^{t} (M_t^* L_t^* C_{RO})$$
$$+ \sum_{i=1}^{m}\sum_{j=1}^{n} (IS_{ij}t + S_{ij}t - D_{ij}t) * C_{IO} \leq C_t$$

The subsequent equation formalizes the revenue objective for cement products, with g_t representing the sales target variable:

$$Z_2t = \sum_{i=1}^{m}\sum_{j=1}^{n} P_{ij}t^* S_{ij}t \geq g_t.$$

The problem constraints are formulated as follows:

Z.t:.

To ensure that the regular working hours of the factory workforce neither fall below the minimum regular working hours nor exceed the maximum regular working hours, we can express this as:

$$N_t \ \min \leq N_t \leq N_t \ \max.$$

The remaining constraints are formulated as follows:

The overtime workforce is constrained by the total employable workers to prevent operational overextension:

$$L_t \leq L_t \ \max.$$

The overtime constraint enforces to balance worker welfare and production needs:

$$M_t \ \min \leq M_t \leq M_t \ \max.$$

The sum of the product inventory from the previous month plus the production quantity in the present month, minus the present month's inventory, must be greater than or equal to the forecasted demand for that product in the present month:

$$S_{ij}t + (IS_{ij}t - 1) - IS_{ij}t \geq D_{ij}t.$$

The crusher capacity constraint is expressed as follows:

$$PC1_{ij}t \leq C.cap1_{ij}t.$$

The mold capacity constraint ensures that the production quantity does not exceed the maximum allowable mold capacity per period:

$$PC2_{ij}t \leq C.cap2_{ij}t.$$

The raw material warehouse capacity constraint ensures that the stored quantity of materials does not exceed the maximum storage capacity:

$$IPt \leq I.cap\ t.$$

The concrete production quantity must not exceed the maximum mold capacity for product manufacturing:

$$RM_j t \leq RM.cap_j\ \ t.$$

The amount of material stored per period must not exceed the maximum capacity of the vibrating table:

$$RS\ t \leq RS.cap\ \ t.$$

The quantity of manufactured products must not exceed the maximum mold capacity:

$$K_{ij}t \leq K.cap_{ij}t.$$

The quantity of additives stored by equipment in the intermediate products warehouse during each period must not exceed the maximum warehouse capacity:

$$10t \leq 10.cap\ t.$$

The concrete vibration capacity for product manufacturing must not exceed the maximum vibration table limits:

$$CM_{ij}t \leq CM.cap_{ij}t.$$

The stored quantity in each concrete silo must not exceed its maximum capacity during any production period:

$$IL\ \ t \leq IL.cap\ \ t$$

$$IM\ \ t \leq IM.cap\ \ t$$

$$IS\ \ t \leq IS.cap\ \ t.$$

The produced concrete must comply with grade specifications, and its quantity must not exceed the maximum capacity for each designated grade:

$$S \leq S.cap_{ij}\ t.$$

The quantity of products manufactured by different machines at Station 1 must equal the quantity of materials entering the raw material warehouse, where m and h denote the number of concrete production machines and vibrating tables, respectively:

$$\sum_{i=1}^{m} \sum_{j=1}^{n} PC_{1ij}t + \sum_{i=1}^{h} \sum_{j=1}^{n} PC_{2ij}t = IPt.$$

The quantity of cement input to the concrete mixer must equal the quantity of material available in the raw material warehouse:

$$\sum_{j=1}^{n} RM_j t = IPt.$$

The input material quantity at the raw material silo stage equals the output of the raw material mill minus the moisture loss during the milling process. Primary drying occurs at this stage, and the production quantity must not exceed the maximum mold capacity:

$$\%X \sum_{j=1}^{n} RM_j t = RSt.$$

The total powdered material stored in the raw material silo equals the material input to the concrete mixer, meaning all material is transferred from the silo to the mixer:

$$\sum_{i=1}^{m} \sum_{j=1}^{n} K_{ij}t = RSt.$$

A percentage of additives is blended with the cement output from the device to produce concrete in the next stage:

$$\%A[10\ t] + (1 - \%A)(ZIt) = \sum_{i=1}^{m} \sum_{j=1}^{n} CM_{ij}t.$$

The following equation specifies that not all input material to the device is converted to concrete, as the transformation depends on the station's efficiency (% y); thus, the material output from the device to the next stage differs from the input quantity:

$$\%y \sum_{i=1}^{m} \sum_{j=1}^{n} K_{ij}t = 10\,t.$$

According to the following three constraints, all produced products are transferred to and stored in the warehouse:

$$\sum_{i=1}^{m-2} \sum_{j=1}^{n} CM_{ij}t = IL\ \ t$$

$$\sum_{i=m-1}^{m-2} \sum_{j=1}^{n} CM_{ij}t = IM\ \ t.$$

$$\sum_{i=1}^{m} \sum_{j=1}^{n} CM_{ij}t = IS\ \ t$$

All produced products must be transferred to the sales station:

$$IL\,t + IM\,t + IS\,t = \sum_{i=1}^{m} \sum_{j=1}^{n} S_{ij}t.$$

References

Abdali, H., Sahebi, H., & Pishvaee, M. (2021). The water-energy-food-land nexus at the sugarcane-to-bioenergy supply chain: A sustainable network design model. *Computers and Chemical Engineering, 145*, 107–122.

Amirteimoori, A., & Kordrostami, S. (2011). Production planning: A DEA-based approach. *The International Journal of Advanced Manufacturing Technology, 56*(1–4), 369–376.

Amirteimoori, A., & Kordrostami, S. (2012). Production planning in data envelopment analysis. *International Journal of Production Economics, 140*(1), 212–218.

Amirteimoori, A. R., & Soufi, M. (2019). Provide a new targeting model in a centralized decision-making environment with a multi-component network structure. *Journal of Operational Research and Its Applications, 16*(1), 93–115.

Asgarpour, M. J. (2023). *Multiple criteria decision making*. University of Tehran Press.

Athanassopoulos, A. D. (1998). Decision support for target-based resource allocation of public services in multiunit and multilevel systems. *Management Science, 44*(2), 173–187.

Athanassopoulos, A. D., Lambroukos, N., & Seiford, L. (1999). Data envelopment scenario analysis for setting targets to electricity generating plants. *European Journal of Operational Research, 115*(3), 413–428.

Charnes, A., & Cooper, W. W. (1961). *Management models and industrial applications of linear programming* (Vol. 1). John Willey.

Cohon, J. L., & Marks, D. H. (1975). A review and evaluation of multi objective programing techniques. *Water Resources Research, 11*(2), 208–220.

Cook, W. D., & Zhu, J. (Eds.). (2010). *Modeling Data Irregularities and Structural Complexities in Data Envelopment Analysis*. Springer.

Du, J., & Liang, L. (2009). Output planning based on data envelopment analysis and extended Cournot model. In *2009 IEEE International Conference on Systems, Man and Cybernetics* (pp. 5117–5122). IEEE.

Du, J., Liang, L., Chen, Y., & Bi, G. B. (2010). DEA-based production planning. *Omega, 38*(1–2), 105–112.

Duckstein, L., & Sudarovszky, F. (1994). Distance base techniques in multi criterion decision making. In J. J. Bogardi & H. P. Nachtnebel (Eds.), *Multi criterion decision analysis in water resources management*. UNESCO.

Fuller, R., & Carlsson, C. (1996). Fuzzy multiple criteria decision making: Recent developments. *Fuzzy Sets and Systems, 78*, 139–153.

Golany, B. (1988). An interactive MOLP procedure for the extension of DEA to effectiveness analysis. *Journal of the Operational Research Society, 39*(8), 725–734.

Hamies, Y. Y., & Hall, W. A. (1974). Multi-objectives in water resources systems analysis: The surrogate worth trade-off method. *Water Resources Research, 10*(4), 615–625.

Hosseinzadeh Lotfi, F., & Moghtaderi, S. (2010). DEA-based production planning changes in general situation. *Applied Mathematical Sciences, 4*(69–72), 3523–3536.

Hwang, C., & Yoon, K. (1981). *Multiple attribute decision making—Methods and applications, a state of the art survey*. Springer-Verlag.

Kazemi Matin, R., & Azizi, R. (2011). A two-phase approach for setting targets in DEA with negative data. *Applied Mathematical Modelling, 35*(12), 5794–5803.

Keeney, R. L., & Raiffa, H. (1993). *Decision with multiple objectives*. Cambridge University Press.

Legesse, G. S., Cattrysse, D., Alemayehu, E., & Orshoven, J. V. (2021). Multi - criteria decision making methods to address rural land allocation problems: A systematic review. *International Soil and Water Conservation Research, 9*(4), 490–501.

Li, H., Yang, W., Zhou, Z., & Huang, C. (2013). Resource allocation models' construction for the reduction of undesirable outputs based on DEA methods. *Mathematical and Computer Modelling, 58*(5–6), 913–926.

Lin, R. (2011). Allocating fixed costs or resources and setting targets via data envelopment analysis. *Applied Mathematics and Computation, 217*(13), 6349–6358.

Lozano, S., & Gutiérrez, E. (2011). A multiobjective approach to fleet, fuel and operating cost efficiency of European airlines. *Computers & Industrial Engineering, 61*(3), 473–481.

Lozano, S., & Villa, G. (2009). Multiobjective target setting in data envelopment analysis using AHP. *Computers & Operations Research, 36*(2), 549–564.

Malekmohammadi, N., Hosseinzadeh Lotfi, F., & Jaafar, A. B. (2009). Centralized resource allocation in DEA with interval data: An application to commercial banks in Malaysia. *International Journal of Mathematical Analysis, 3*(13–16), 757–764.

Mehregan, M. R. (2013). *Multiple objective decision making*. University of Tehran Press.

Thanassoulis, E., & Dyson, R. G. (1992). Estimating preferred target input-output levels using data envelopment analysis. *European Journal of Operational Research, 56*(1), 80–97.

Triantis, K. P., & Liu, S. (2011). Manufacturing performance measurement and target setting: A data envelopment analysis approach. *European Journal of Operational Research, 214*(3), 616–626.

Yuan, G. (2012). Two-stage fuzzy production planning expected value model and its approximation method. *Applied Mathematical Modelling, 36*(6), 2429–2445.

Zeleny, M. (1973). Compromise programming. In J. L. Cachrone & M. Zeleny (Eds.), *Multiple criteria decision making*. University of South Carolina Press.

Zahraee, S. M., Shiwakoti, N., & Stasinopoulos, P. (2020). Biomass supply chain environmental and socio-economic analysis: 40-years comprehensive review of methods, decision issues, sustainability challenges, and the way forward. *Biomass and Bioenergy, 142*, 105777.

Chapter 2
Pairwise Comparison Methods

Learning Objectives

Upon completion of this chapter, the reader will be able to:

1. **Explain** the fundamental principles, advantages, and disadvantages of pairwise comparison as a foundational technique in multi-criteria decision-making (MCDM).
2. **Differentiate** between key pairwise comparison methods, including the analytic hierarchy process (AHP), the analytic network process (ANP), and the best-worst method (BWM), and identify the contexts for which each is most suitable.
3. **Construct** pairwise comparison matrices, calculate priority weights using methods such as the eigenvector technique, and **assess** the consistency of judgments using the consistency ratio (CR).
4. **Apply** the DEMATEL (Decision-Making Trial and Evaluation Laboratory) method to analyze cause-effect relationships among factors in a complex system and visualize them through an impact-relations map.
5. **Understand** the rationale for integrating fuzzy logic with MCDM techniques and **describe** the hybrid fuzzy DEMATEL-based ANP (DANP) approach for handling uncertainty and interdependencies in real-world problems.
6. **Implement** the step-by-step methodology of the best-worst method (BWM) to determine criteria weights with reduced pairwise comparisons and evaluate the consistency of the results.

A. Amirteimoori, M. Soufi, *Industrial Decision Analysis*, Multiple Criteria Decision Making, https://doi.org/10.1007/978-3-032-24299-0_2

2.1 Introduction

Pairwise comparison methods in multi-criteria decision-making (MCDM) constitute a category of techniques where alternatives (or criteria) are compared in pairs. These comparisons are typically based on the judgments and opinions of the decision-maker, and their outcome determines the degree of preference of one alternative over another. These methods are particularly useful when the number of alternatives is not large, and a direct comparison between them is feasible (Amirteimoori & Soufi, 2019).

The fundamental idea in pairwise comparison methods is that instead of attempting to rank all alternatives simultaneously, they are compared two at a time, and then a final ranking is established using the results of these comparisons. This process is often simpler and more intuitive than evaluating all alternatives concurrently (Abdali et al., 2021).

Advantages of Pairwise Comparison Methods

- **Ease of Understanding and Implementation:** Comparing two alternatives is generally easier for individuals than comparing several alternatives simultaneously.
- **Reduced Cognitive Complexity:** By dividing the problem into smaller sub-problems, this method reduces the cognitive complexity for the decision-maker.
- **Applicability in Various Contexts:** These methods can be used in various situations, including individual and group decision-making.
- **Capability of Quantifying Qualitative Judgments:** Qualitative judgments of the decision-maker regarding the preference between alternatives can be expressed quantitatively (e.g., using numerical scales).

Disadvantages of Pairwise Comparison Methods

- **Increased Number of Comparisons with Increasing Alternatives:** The number of comparisons required increases exponentially with the number of alternatives $\frac{n(n-1)}{2}$, where n is the number of alternatives. This can be time-consuming and tedious in problems with a large number of alternatives.
- **Potential for Inconsistencies in Judgments:** The decision-maker's judgments in different comparisons may not be consistent with each other (e.g., if alternative A is preferred over B, and B is preferred over C, it is expected that A is also preferred over C, but this may not be the case).

Prominent Pairwise Comparison Methods

1. **Analytic Hierarchy Process (AHP)**

    - This method is one of the most widely used MCDM methods, developed by Thomas Saaty (Mikhailov, 2003).
    - In AHP, the decision problem is modeled as a hierarchical structure consisting of the objective, criteria, and alternatives.
    - Then, using pairwise comparisons, the weights of the criteria and the scores of the alternatives on each criterion are determined.

- Finally, by combining these weights and scores, the final ranking of the alternatives is obtained.

2. **Analytic Network Process (ANP)**
   - ANP is a generalization of AHP, where the interdependencies between criteria and alternatives are also considered.
   - AHP assumes that the criteria are independent of each other, whereas ANP removes this assumption and allows for interdependencies and feedback relationships between criteria.

Conducting Pairwise Comparisons
A numerical scale is typically used to conduct pairwise comparisons, representing the degree of preference of one alternative over another. Saaty's scale is one of the most common scales, as follows:

- 1: Equal preference
- 3: Moderate preference
- 5: Strong preference
- 7: Very strong preference
- 9: Extreme preference
- 2, 4, 6, 8: Intermediate values (Asgarpour, 2023)

Simple Example
Assume we want to choose between three alternatives A, B, and C. Using pairwise comparisons, we make the following judgments:

- A is moderately preferred over B (3).
- B is strongly preferred over C (5).
- A is very strongly preferred over C (7).

  Using these judgments and AHP or ANP techniques, the final ranking of the alternatives can be obtained.

  In summary, pairwise comparison methods are powerful tools for decision-making in multi-criteria problems, especially when the number of alternatives is not large and there is a need to consider the decision-maker's qualitative judgments.

2.2 What Is Pairwise Comparison?

Pairwise comparison is the process of comparing a set of options using head-to-head pairs to judge which one is the most preferred overall. Also known as "pairwise ranking," it is a popular research method used for ranking people's preferences, informing strategic decisions, and conducting voting at scale (Hwang & Yoon, 1981).

Pairwise comparison works by breaking a set of options down into a series of head-to-head votes. Once a respondent picks their preference from the two options, their vote is recorded and a new pair of options from the set are shown for the next vote.

Pairwise comparison is a popular survey question format used for ranking a list of options. For example, consider a scenario where you need to select one employee for a departmental head position, and a list of 20 employees has been provided (Hamies & Hall, 1974). There is no readily measurable way to rank a subjective list such as a group of individuals; therefore, attempting to rank all of them simultaneously in an ordered list is exceedingly difficult. Instead, breaking the names into several pairs and selecting the preferred individual from each pair only takes a few minutes and generates a list of individuals ranked from most to least preferred (Duckstein & Sudarovszky, 1994).

In this method, each candidate is compared pairwise with the remaining candidates. The number of times a candidate is preferred over others is recorded as their "wins." This process is repeated for all candidates, and ultimately selection is based on the total number of wins or preferences received.

In a pairwise comparison survey, one respondent can be tasked with voting on every possible pair combination, or a bunch of people can each be given a sample of pairs to complete that can later be used to calculate the group's overall combined preferences (Fuller & Carlsson, 1996).

The total number of possible pairwise comparisons from a list of options is $\frac{n(n-1)}{2}$, where "n" stands for the number of options in the set. For example, a set of ten options has 45 possible pairwise comparisons ($\frac{10(10-1)}{2}$).

2.2.1 Methods for Calculating Pairwise Comparison Results

There are three calculation methods we can use when scoring pairwise comparison voting data:

1. **The Win Rate Method**

   The win rate method in pairwise comparison is a straightforward and widely used technique for ranking alternatives based on the results of their pairwise comparisons. Rather than directly ranking the alternatives, they are compared in pairs, and the number of times each alternative is preferred over others is tallied (Hwang & Yoon, 1981).

Calculation Steps for Win Rate

1. **Pair Formation:** Initially, all possible pairs of alternatives are formed. For instance, given four alternatives (A, B, C, and D), the resulting pairs are (A, B), (A, C), (A, D), (B, C), (B, D), and (C, D).
2. **Pair Comparison:** Respondents are asked to select their preferred alternative within each pair.
3. **Win Count:** For each alternative, the number of times it "wins" in the pairwise comparisons (i.e., is preferred over the other alternative) is counted.
4. **Win Rate Calculation:** The win rate for each alternative is calculated using the following formula:

Win rate = (number of wins/total number of comparisons the alternative appeared in) * 100.

Example 2.1 Consider four alternatives (A, B, C, and D) with the following pairwise comparison results:

- A vs. B: A wins.
- A vs. C: C wins.
- A vs. D: A wins.
- B vs. C: B wins.
- B vs. D: D wins.
- C vs. D: C wins.

The win rate for each alternative is calculated as follows:

- Alternative A: 2 wins out of 3 comparisons => win rate = $\left(\frac{2}{3}\right) \times 100 = 66.67\%$.
- Alternative B: 1 win out of 3 comparisons => win rate = $\left(\frac{1}{3}\right) \times 100 = 33.33\%$.
- Alternative C: 2 wins out of 3 comparisons => win rate = $\left(\frac{2}{3}\right) \times 100 = 66.67\%$.
- Alternative D: 1 win out of 3 comparisons => win rate = $\left(\frac{1}{3}\right) \times 100 = 33.33\%$.

Therefore, alternatives A and C, with a win rate of 66.67%, are ranked first, while alternatives B and D, with a win rate of 33.33%, are ranked lower.

Key Considerations

- The win rate is easily understood and calculated, making it a widely applicable method.
- The win rate represents the probability of an alternative winning if compared randomly against another alternative from the same set.
- When the number of alternatives is large, the number of pairwise comparisons increases significantly. In such cases, the use of software tools can be beneficial.
- If two or more alternatives have the same win rate, supplementary criteria can be used to determine their ranking. For example, the number of times each alternative was chosen as the "first choice" in the comparisons could be considered.

2. **Probabilistic Methods**

Probabilistic methods in pairwise comparisons offer a distinct approach from the win rate method. Instead of directly counting wins and calculating percentages, these methods employ probabilistic models and statistical algorithms, particularly Bayesian algorithms, to analyze patterns of preference within pairwise comparison data. The primary objective of these methods is to estimate the strength or "skill" of each alternative relative to others, with these estimations being updated and refined over time as more data is acquired (Keeney & Raiffa, 1993).

Key Characteristics of Probabilistic Methods

- **Probabilistic Modeling:** These methods are based on assumptions about the probabilistic distribution of individual preferences. In other words, they assume that selecting one alternative over another is a stochastic event with a specific probability.
- **Dynamic Updating:** Scores (or rankings) in these methods are not static; they are adjusted after each new comparison based on the outcome. This feature makes these methods suitable for environments where preferences may change over time.
- **Baseline Starting Point:** These models typically begin with an initial score for all alternatives (such as 1500 in the ELO system), which are then modified based on the results of the comparisons.
- **Increased Complexity:** Compared to the win rate method, probabilistic methods are generally more computationally complex and require a greater understanding of statistical concepts for comprehension and implementation.

Prominent Probabilistic Algorithms in Pairwise Comparisons

- **ELO[1]:** Primarily known for its use in ranking chess players, this algorithm is one of the simpler yet widely used probabilistic methods. In ELO, a player with a higher rating is expected to win against a player with a lower rating. The magnitude of score changes after each match depends on the rating difference between the two players and the match outcome.
- **Glicko:** This algorithm is an extension of ELO that also calculates a "rating deviation" (RD). RD represents the uncertainty about a player's rating. A higher RD indicates greater uncertainty, leading to larger rating changes after matches.
- **TrueSkill:** Developed by Microsoft, this algorithm is designed for multiplayer games. Instead of a single rating, TrueSkill considers a probability distribution for each player's "skill" and can simultaneously update the ratings of multiple players in a single match.

Advantages and Disadvantages of Probabilistic Methods

Advantages

Higher Accuracy (in Some Cases): In situations with noisy data or where preferences change over time, probabilistic methods can provide more accurate rankings than the win rate method.

Modeling Uncertainty: These methods can also indicate the level of uncertainty associated with the rankings.

---

[1] ELO is the surname of its creator, Arpad Elo, a Hungarian-American physicist.

Suitability for Dynamic Environments: The dynamic updating capability makes these methods suitable for environments with evolving preferences.

Disadvantages

Computational Complexity: These methods are typically more difficult to understand and implement compared to the win rate method.

More Difficult Interpretation: Directly interpreting scores in these methods can be challenging, especially for those unfamiliar with statistical concepts.

Higher Data Requirements (in Some Cases): Achieving accurate estimations may require more data compared to the win rate method.

Applications

Besides video games and chess, probabilistic methods are also used in other areas, such as:

- **Recommender systems:** For suggesting products or services to users based on their preferences
- **Performance evaluation:** For evaluating the performance of individuals or teams in competitive environments
- **Surveys and market research:** For more detailed analysis of preference data

Ultimately, the choice between the win rate method and probabilistic methods depends on the specific context and analytical objectives. If simplicity and ease of interpretation are important, the win rate method is a suitable option. However, if higher accuracy, uncertainty modeling, and dynamic updating are required, probabilistic methods may be a better choice.

3. **The Manual Method**

The manual method for calculating pairwise comparison results is a simple way to rank options based on preferences, provided that the number of options and the number of voters are limited (typically one voter). This method is based on constructing a pairwise comparison matrix (Hwang & Yoon, 1981). The steps are explained below:

1. *Creating the Pairwise Comparison Matrix*
   - Draw a square matrix.
   - Write the options in both the rows and columns of the table. The order of options should be the same in rows and columns.

Example 2.2 Suppose we Want to Compare Three Options a, B, and C. The Matrix Will Look like this

	A	B	C
A	–		
B		–	
C			–

2. *Performing Comparisons*

   - Start from the first row and compare each option in the row with each option in the columns.
   - For each comparison, decide which option is preferred.
   - If the row option is preferred over the column option, put 1 in the corresponding cell.
   - If the column option is preferred over the row option, put 0 in the corresponding cell.
   - In the cells where an option is compared with itself (the main diagonal of the matrix), usually nothing is placed or a dash (–) is used.

   Example (continued): Suppose the preferences are as follows:

- A is preferred over B.
- A is preferred over C.
- B is preferred over C.

The completed matrix will look like this:

3. *Calculating Scores.*

   - For each option, sum the numbers in the row corresponding to that option. This sum will be the score of that option.
   - Score of A: 1 + 1 = 2
   - Score of B: 0 + 1 = 1
   - Score of C: 0 + 0 = 0

4. *Ranking Options.*

   - Rank the options based on their scores from highest to lowest.

   The ranking will be as follows:

1. A (Score 2)
2. B (Score 1)
3. C (Score 0)

If two options have the same score, both options can be given the same rank, or other methods can be used to break the tie (such as a closer examination of the comparison results). Some versions give 0.5 to both options in case of a tie in the comparison of two options.

	A	B	C
A	–	1	1
B	0	–	1
C	0	0	–

2.3 Analytic Hierarchy Process (AHP)

This process is a multiple criteria decision-making (MCDM) method employed to select the optimal alternative from a set of options. The decision-maker determines the criteria used to evaluate these alternatives.

Analytical hierarchy process (AHP) is a decision-making framework that incorporates both subjective and objective assessments to address complex problems with multiple, often conflicting, criteria. By breaking down the problem into a hierarchical structure, AHP facilitates a systematic evaluation of alternatives against various objectives.

The implementation of AHP can be structured into either eight or four distinct steps. While Lockwood's approach outlines eight steps, other sources have simplified the process into four key stages: decomposition, weighting, evaluation, and selection.

Various methods have been proposed to determine the weights assigned to each criterion:

1. The eigenvector
2. The averaged least squares method (ALSM)
3. The logarithmic least squares method (LLSM)
4. The weighted least squares method (WLSM)
5. The geometric row mean (GRM)
6. Several methods based solely on mathematical calculations, including the normalized matrix row average method and the normalized total rows and total columns reversed method

1. *The eigenvector method*, pioneered by Saaty (1980), is a widely recognized and frequently employed technique. This method derives its solution from the eigenvector, and its foundation in the mini-max principle has been established by Frobenius (Sekitani & Yamaki, 1999; Sekitani, 2000). Initially confined to AHP applications, the eigenvector method has since expanded its scope to encompass general mutual evaluation network (MEN) problems.

The eigenvector method proceeds as follows:

$$W_i = \sum \frac{S_{ij} W_j}{n_i} \quad (i = 1, 2, \ldots., n).$$

2. *The ALS method* is an iterative procedure designed to minimize the sum of squared deviations (d_i) between a set of estimated values and their respective target values (W_i). This method makes $\overline{W}_i$ close to W_i by minimizing the sum of the squares of the $d_i = \overline{W}_i - W_i$:

$$d_i = \overline{W}_i - W_i = \sum \frac{S_{ij} W_j}{n_i}$$

$$\begin{aligned} &Min\, D = \sum d_i^2 \\ &W_1 + W_2 + \ldots. + W_n = 1 \end{aligned}.$$

3. *The LLS method* is well known. Taking the logarithm of the above equations yields:

$$W_t - W_j = S_{ij} \qquad \left(W_t = \ln W_i - S_{tj} = \ln S_{ij}\right)$$

$$\begin{aligned} &Min\, D = \sum \left[\ln S_{ij} - \left(\ln W_i - W_j\right)\right]^2 \\ &\ln W_1 + \ln W_2 + \ldots. + \ln W_n = 0 \end{aligned}.$$

The LLS method is a straightforward approach, and for AHP problems with complete information, its solution is equivalent to that of the geometric mean method. This ensures a positive solution, as the exponential transformation of logarithmic values always yields positive results.

4. *The WLS method* is a straightforward approach. Its principle is relatively simple:

$$e_{ij} = W_i - S_{ij} W_j$$

$$Min\, D = \sum\nolimits_{(j,i) \in E} e_{ij}^2 = \sum\nolimits_{(j,i) \in E} \left(W_i - S_{ij} W_j\right)^2.$$

Nishizawa and Takahashi (2008) reviewed five primary methods for mutual evaluation network systems, suggesting that few, if any, other significant methods exist. The geometric mean method is equivalent to the LLS method for AHP problems with complete information. The direct least squares method may yield multiple solutions, necessitating nonlinear computational techniques, rendering it less practical. Nishizawa and Takahashi advocate for the superiority of the LLS and WLS methods.

Table 2.1 Effective factors of establishing e-factory system for management and human resource

Criteria	Effective elements of implementing e-factory at Iran industries
Management	1. Lack of managers' mastery at economic and international strategic problems 2. Ruling of national vision instead of global vision in managers 3. Inattention to capitals' arrival and foreign capitalist to industry 4. Lack of integrated information system of management at Iran's industries 5. Rapid replacement of top managers due to their belonging to government 6. Lack of strategies for keeping valuable human resources
Human resource	1. Lack of skilled workforce 2. Lack of flexibility of personnel against changes 3. Lack of full awareness of personnel with description of their duties 4. Lack of personnel's commitment for achieving organization's goal

2.3.1 Case Study

The Role of Management and Human Resource in Implementing e-Factory by AHP Method (Case Study of Guilan Province Industries) (Ahmad Jafarnejad et al., 2014)

To illustrate the decision-making process using the analytic hierarchy process (AHP), a research project was conducted to investigate the barriers to implementing an electronic factory. The primary objective of this research was to identify and rank all factors influencing the implementation of an electronic factory. Data was collected through a limited distribution of specialized questionnaires among industry experts.

The research methodology comprised three main stages. In the initial stage, factors confronting industries in establishing an electronic factory were identified based on the questionnaires. Cronbach's alpha coefficient was employed to assess the reliability of the indicators. Subsequently, the identified factors were prioritized through the construction of an AHP model using the Expert Choice software. In the final stage, based on the prioritized barriers, appropriate procedures were proposed to address the internal and external environmental conditions.

First Step: In this study, through a comprehensive review of the theoretical literature on the development of electronic factories, the dimensions and indicators influencing the establishment of such facilities were identified. Subsequently, a specialized questionnaire (Questionnaire A) was designed and distributed to experts from various industries to assess and prioritize the indicators and factors affecting the implementation of electronic factory systems (Table 2.1). Cronbach's alpha coefficient was employed to evaluate the reliability of the questionnaire items.

Second Step: An aggregate pairwise comparison matrix is constructed by aggregating the responses from industry experts who have completed the pairwise comparison questionnaires. The geometric mean is typically employed for this purpose.

Aggregate pairwise comparison matrices were constructed for each group of elements at different levels of the AHP hierarchy. These matrices were calculated

Table 2.2 Composed matrix of paired comparisons of management elements

	1	2	3	4	5	6
1	**1**	0.38	2.55	1.93	0.71	1.85
2	0.42	**1**	1.61	0.84	0.65	0.90
3	0.39	0.62	**1**	1.05	0.54	0.74
4	0.52	1.19	0.95	**1**	0.65	0.91
5	1.41	1.54	1.86	1.54	**1**	1.20
6	0.54	1.11	1.36	1.11	0.83	**1**

Table 2.3 Composed matrix of paired comparisons of human resource elements

	1	2	3	4
1	**1**	2.33	0.65	0.79
2	0.43	**1**	0.47	0.61
3	1.54	2.13	**1**	1.18
4	1.26	1.63	0.84	**1**

separately for each element. Table 2.2 presents the aggregated pairwise comparison matrix for the management elements (Table 2.3).

Data Normalization in Multi-Criteria Decision-Making (MCDM)
In multi-criteria decision-making (MCDM) methods, we often deal with data that have different units of measurement. For instance, when selecting a car, we might consider criteria such as price, fuel consumption, and acceleration. Price is measured in dollars, fuel consumption in liters per kilometer, and acceleration in meters per second squared. How can we compare these criteria? This is where data normalization comes into play.

Normalization is a process that transforms data to a common scale, enabling comparison between different criteria. This is especially crucial in MCDM methods as we deal with multiple criteria that may have varying units of measurement.

Why Is Normalization Important?

- Differences in Units of Measurement: As illustrated in the car example, different criteria often have different units of measurement. Without normalization, direct comparison between these criteria is not feasible.
- Range of Values: The values of each criterion may vary significantly. This difference in range can cause criteria with larger values to have a greater influence on the decision-making process, even if their inherent importance is lower.

Normalization Methods
Several normalization methods exist, each with its own advantages and disadvantages. Some common methods include:

1. Linear normalization:

    The simplest method. Each value is divided by the largest value in that criterion.

    Suitable for positive criteria (larger is better):

$$Formula : Normalized\ value = \frac{Original\ value}{Largest\ value\ in\ that\ criterion}$$

2. Vector normalization:

   Based on Euclidean distance. Suitable for cases where the distance between options is important:

$$Formula : Normalized\ value = \frac{Original\ value}{\sqrt{Sum\ of\ squares\ of\ all\ values\ in\ that\ criterion}}$$

3. Rumina normalization:

   Suitable for both positive and negative criteria. For positive criteria, it's similar to linear normalization. For negative criteria, the smallest value is divided by the original value.
4. Standardization:

   Subtracts the mean of a column from each value and then divides by the standard deviation of that column.

Standardized values have a mean of zero and a standard deviation of one.

Choosing the Appropriate Method

The choice of normalization method depends on various factors, including:

Type of criterion: Positive, negative, or mixed
Relative importance of criteria
MCDM method used

Importance of Normalization in MCDM

Normalization is a crucial step in many MCDM methods such as AHP, TOPSIS, ELECTRE, and others. Without normalization, the results obtained from these methods would not be reliable.

Data normalization is a vital step in multi-criteria decision-making. By normalizing data, we can convert criteria with different units of measurement to a common scale, enabling fair comparison between alternatives. The choice of normalization method depends on the specific problem and the MCDM method employed.

Sum normalization is employed in this study. Sum normalization, also known as percentage normalization, is a technique where each value in a dataset is expressed as a percentage of the total sum. This method allows us to determine the relative importance of each value compared to others within the dataset. In this method, each element is divided by the sum of the elements in its column. By doing so, we express each value as a percentage of the column total, allowing us to determine the relative contribution of each element within the column (Tables 2.4, 2.5, 2.6, 2.7, and 2.8).

The relative weight assigned to each indicator is calculated based on the arithmetic mean of the values in the corresponding row, which is represented by the notation W_i.

Table 2.4 Normalized matrix of management elements on implementing e-factory system

	1	2	3	4	5	6	W_i
1	0.234	0.065	0.273	0.258	0.162	0.280	**0.212**
2	0.098	0.171	0.173	0.112	0.148	0.136	**0.140**
3	0.091	0.106	0.107	0.141	0.123	0.112	**0.113**
4	0.121	0.204	0.102	0.134	0.148	0.138	**0.141**
5	0.329	0.264	0.199	0.206	0.228	0.182	**0.235**
6	0.126	0.190	0.146	0.149	0.189	0.152	**0.159**

Table 2.5 Normalized matrix of human resource elements on implementing e-factory system

	1	2	3	4	Wi
1	0.236	0.329	0.220	0.221	**0.251**
2	0.102	0.141	0.159	0.170	**0.143**
3	0.364	0.300	0.338	0.330	**0.333**
4	0.298	0.230	0.284	0.279	**0.273**

Table 2.6 Calculate the weighted sum vector and the consistency vector

WSV management	CV management	WSV human resource	CV human resource
1.528	6.112	1.015	4.0398
1.805	6.053	0.574	4.0397
0.657	5.521	1.436	4.0395
0.799	5.963	1.104	4.0395
1.365	6.066		
0.909	6.992		
Sum	36.707		16.1558 .

Table 2.7 Random index

n (matrix size)	1	2	3	4	5	6	7	8	9	10
RI (random index)	0	0	0.58	0.9	1.12	1.24	1.32	1.41	1.45	1.49

Table 2.8 Calculate the consistency ratio

	λmax	C1	RI	CR
Management	6.118	0.0236	1.24	0.019
Human resource	4.0396	0.0099	0.9	0.011

The rate of inconsistency of composed matrices of paired comparison is calculated as follows. For example, the calculation steps for the matrix related to "management agent" are as follows.

What Is the Inconsistency Ratio in the Analytic Hierarchy Process (AHP)?
The inconsistency ratio (IR) is a critical measure employed in the analytic hierarchy process (AHP) to assess the consistency and reliability of judgments made by experts during pairwise comparisons. In simpler terms, the IR indicates the degree to which the judgments regarding the relative importance of criteria and alternatives are congruent.

Why Is the Inconsistency Ratio Important?

Validity of Results: A high inconsistency ratio suggests that the judgments are contradictory and, consequently, the results obtained from the AHP analysis may be unreliable.

Refinement of Judgments: An elevated inconsistency ratio indicates a need to revisit and refine the judgments to reduce inconsistencies.

Calculating the Inconsistency Ratio

The precise calculation of the inconsistency ratio requires specialized knowledge of AHP and is typically performed using dedicated software. However, in general, the process involves computing the consistency index (CI) and comparing it to a random index (RI).

Consistency Index (CI): This index quantifies the level of inconsistency present in the pairwise comparison matrix.

Random Index (RI): The RI represents the average consistency index for randomly generated matrices of the same size.

Inconsistency Ratio = CI/RI

If the inconsistency ratio is less than 0.1, the pairwise comparison matrix is considered acceptably consistent.

Factors Influencing Inconsistency Ratio

Number of Criteria and Alternatives: A larger number of elements can increase the likelihood of inconsistencies.

Expert Experience and Knowledge: Experts with greater experience and knowledge tend to provide more consistent judgments.

Clarity of Criteria Definitions: Ambiguous or poorly defined criteria can contribute to higher inconsistency levels.

Reducing Inconsistency

Revising Judgments: A high inconsistency ratio necessitates a careful review and potential modification of the judgments.

Increasing the Number of Experts: Involving more experts can introduce diverse perspectives and help mitigate inconsistencies.

Utilizing Specialized Software: AHP software can aid in identifying and addressing inconsistencies.

Suitable Software for AHP Analysis

Commercial Software

Expert Choice: This software is one of the oldest and most well-known AHP tools. Expert Choice offers extensive capabilities for creating complex AHP models, calculating weights, and conducting sensitivity analysis.

Super Decisions: Another software used for both AHP and ANP (analytic network process) analysis, Super Decisions provides advanced features for modeling and analyzing complex systems.

CPLEX: A powerful optimization solver that can also be used for solving AHP problems.

MATLAB: With its various functions and toolboxes, MATLAB can be used for AHP analysis.

Open-Source Software

R: The R programming language, using packages like "hierarchy" and "AHP," enables AHP analysis.

Python: The Python programming language, with libraries such as "NumPy" and "SciPy," can be employed for AHP analysis.

Spreadsheet Software

Microsoft Excel: With add-ins and macros, simple AHP analysis can be conducted in Excel. However, for large and complex projects, specialized software is recommended.

Factors Influencing Software Selection

Model Complexity: For simple models, Excel or R might suffice, while complex models may require specialized software like Expert Choice or Super Decisions.

Software Capabilities: Each software has its unique features. Consider your specific needs before choosing.

Cost: Commercial software typically has higher costs. If budget is a constraint, open-source options can be explored.

Ease of Use: Some software have simpler user interfaces than others.

Key Considerations

Choosing the Right Software: The selection of software depends on the complexity of the problem, budget, and user experience.

Learning the Software: Before using any software, it is advisable to consult the documentation and tutorials.

Interpreting Results: Even the best software cannot think for the user. Interpreting the results of AHP analysis requires knowledge and experience.

Ultimately, selecting the appropriate software will enable you to conduct AHP analysis accurately and efficiently:

$$W.S.V = AW$$

$$\begin{pmatrix} 1 & 2.38 & 2.55 & 1.93 & 0.71 & 1.85 \\ 0.42 & 1 & 1.61 & 0.84 & 0.65 & 0.90 \\ 0.39 & 0.62 & 1 & 1.05 & 0.54 & 0.74 \\ 0.52 & 0.19 & 0.95 & 1 & 0.65 & 0.91 \\ 1.41 & 1.54 & 1.86 & 1.54 & 1 & 1.20 \\ 0.54 & 1.11 & 1.36 & 1.11 & 0.83 & 1 \end{pmatrix} \times \begin{pmatrix} 0.250 \\ 0.133 \\ 0.119 \\ 0.134 \\ 1.224 \\ 0.130 \end{pmatrix} = \begin{pmatrix} 1.528 \\ 0.805 \\ 0.657 \\ 0.799 \\ 1.365 \\ 0.909 \end{pmatrix}$$

$$C.V = \frac{W.S.V}{W}$$

$$C.V = \begin{pmatrix} 6.112 \\ 6.053 \\ 5.521 \\ 5.962 \\ 6.088 \\ 6.992 \end{pmatrix}$$

$$CI = \frac{\lambda \max - n}{n - 1}$$

$$\lambda \max = \frac{\sum_{i=1}^{6} C.V_i}{n} = \frac{36.707}{6} = 6.118$$

$$CI = \frac{6.118 - 6}{6 - 1} = 0.0236.$$

In the analytic hierarchy process (AHP), a structured technique for dealing with complex decisions, the assessment of consistency within pairwise comparison matrices is crucial. After establishing these matrices and calculating the weights of criteria, it is necessary to evaluate the consistency of the judgments. This is achieved using the consistency index (CI) and the consistency ratio (CR).

To account for the fact that larger matrices naturally tend to have higher CI values even when judgments are random, the random index (RI) is employed for normalization. The RI is derived from a table established by Thomas L. Saaty, the originator of the AHP. This table provides average CI values computed from numerous randomly generated reciprocal matrices of different orders (n). The RI values are tabulated for different matrix sizes (n), typically ranging from 1 to 10 or more.

Therefore, the RI value, specific to the size (n) of the pairwise comparison matrix, is retrieved from this standardized table to calculate the consistency ratio (CR) using the formula: CR = CI/RI. A CR value of 0.1 or less is generally considered acceptable, indicating a reasonable level of consistency in the pairwise comparisons. If the CR exceeds this threshold, the judgments within the matrix are deemed inconsistent, and a review of the comparisons is recommended:

$$CR = \frac{CI}{RI} = \frac{0.0236}{1.24} = 0.019$$

Aggregate matrices derived from pairwise comparisons concerning the management factor exhibit acceptable consistency when the consistency ratio is below 0.1 (Mikhailov, 2003). It should be noted that, owing to the application of the geometric mean, the consistency ratio of collective matrices tends to be lower than that of individual matrices.

Third Step: Having confirmed the accuracy and consistency of the calculations, the total weights of the identified barriers to e-factory system implementation in industries were computed to prioritize them. The Expert Choice (EC) software, by

Table 2.9 Prioritizing management and human resource factors on implementation of e-factory system with AHP method

Row	Prioritizing barriers effective on implementation of e-factory system with AHP method	Total weight
1	Lack of managers' mastery at economic and strategic problems	0.0690
2	Rapid replacement of top managers due to their belonging to government	0.0618
3	Lack of personnel's commitment for achieving organization's goal	0.0552
4	Lack of full awareness of personnel with description of their duties	0.0453
5	Lack of flexibility of personnel against changes	0.0416
6	Lack of integrated information system of management	0.0369
7	Ruling of national vision instead of global vision	0.0367
8	Lack of strategies for keeping valuable human resources	0.0358
9	Inattention to capital's arrival and foreign capitalist to industry	0.0328
10	Lack of skilled and specialized workforce	0.0235

aggregating all matrices corresponding to the hierarchical structure, determined the overall weights based on the established hierarchy, subsequently ranking the barriers from highest to lowest weighted importance. Table 2.9 presents the prioritization of management and human resource factors influencing e-factory system implementation using the AHP method.

2.4 Analytic Network Process (ANP)

The analytic network process (ANP) is a multi-criteria decision-making (MCDM) method employed to determine criteria weights and select the optimal alternative based on pairwise comparisons. It is a generalization of the analytic hierarchy process (AHP) and is applicable in scenarios where criteria and alternatives are not independent and interdependencies and feedback exist among them.

Key Difference Between ANP and AHP

AHP assumes that criteria and sub-criteria are hierarchically structured and independent of each other. In other words, the influence of a criterion is limited to its lower-level sub-criteria, with no reciprocal influence between criteria or between alternatives. However, this assumption often does not hold in real-world situations, where criteria and alternatives can influence each other. ANP addresses this limitation of AHP by considering interdependencies and feedback relationships among decision elements.

Main Steps of ANP

1. Network Modeling: In this step, the decision problem is represented as a network consisting of clusters and elements. Clusters are groups of related elements, and elements represent criteria, sub-criteria, or alternatives. Relationships between elements and clusters are depicted by arrows indicating the direction of influence.

2. Pairwise Comparison Matrix Construction: In this step, elements within each cluster are compared pairwise with respect to other elements within the same cluster or other clusters. These comparisons are based on expert judgment and utilize a ratio scale (typically a nine-point scale). The results of these comparisons are recorded in pairwise comparison matrices.
3. Supermatrix Formation: The supermatrix is a large matrix that contains all the pairwise comparison matrices. This matrix represents the relationships between all elements and clusters.
4. Supermatrix Normalization: The initial supermatrix is usually not normalized. For subsequent calculations, it must be normalized. Normalization is performed such that the sum of the elements in each column equals one.
5. Limit Supermatrix Calculation: By repeatedly multiplying the normalized supermatrix by itself (raising it to a power), the limit supermatrix is obtained. This supermatrix represents the final weights of the elements.
6. Alternative Prioritization: Using the final element weights, alternatives can be prioritized, and the optimal alternative can be selected.

Applications of ANP

ANP is applicable in complex decision-making problems where interdependencies and feedback relationships exist among elements. Some of these applications include:

- Strategic management
- Supplier selection
- Project management
- Risk assessment
- Environmental decision-making

Advantages of ANP over AHP

- Considers interdependencies and feedback relationships among elements
- Provides greater flexibility in modeling complex problems
- Yields more accurate results in cases where the assumptions of AHP are not met

Software Used in ANP

- Super Decisions
- ANP Excel Solver

2.5 DEMATEL

DEMATEL (Decision-Making Trial and Evaluation Laboratory) is a multi-criteria decision-making technique used to analyze cause-and-effect relationships between complex factors within a system. It's particularly useful for understanding the interactions between these factors and identifying key drivers through a visual structural model. DEMATEL was developed by Gabus and Fontela at the Battelle Geneva Research Centre in the early 1970s (specifically around 1971–1976).

Applications
DEMATEL is applied across various fields, including:

- Management and planning: Identifying factors influencing project and program success
- Industrial engineering: Analyzing production systems and pinpointing bottlenecks
- Economics and marketing: Examining factors affecting consumer behavior and market competition
- Social sciences: Analyzing relationships between social and political factors
- Risk assessment: Determining the interdependencies and influences of various risk factors

Main Steps of the DEMATEL Method

1. Define Factors: Identify and define the factors relevant to the system being studied.
2. Construct the Direct Relation Matrix (DRM): Experts assess the pairwise relationships between factors using a defined scale (typically 0–4 or 0–5), where:
    - 0: No influence
    - 1: Low influence
    - 2: Medium influence
    - 3: High influence
    - 4 (or 5): Very high influence

    These assessments are recorded in the DRM.
3. Normalize the Direct Relation Matrix: The DRM is normalized to ensure values fall between 0 and 1. This is done by dividing each element by the maximum sum of either the rows or columns. The normalization formula is:
4. D = M/max (max (sum of rows of M), max (sum of columns of M)).where:
    - D is the normalized direct relation matrix.
    - M is the direct relation matrix.
5. Compute the Total Relation Matrix (TRM): The TRM shows both direct and indirect influences between factors. It's calculated using the following formula:

$$T = D(I - D)^{-1}$$

    where:

    - T is the total relation matrix.
    - D is the normalized direct relation matrix.
    - I is the identity matrix.
    - ^-1 denotes the matrix inverse.

6. Calculate the Row and Column Sums (R and C): From the TRM:
    - Row Sum (R): The sum of each row represents the total influence a factor exerts on other factors (dispatching power).
    - Column Sum (C): The sum of each column represents the total influence a factor receives from other factors (receiving power).
7. Calculate Prominence (R+C) and Net Effect (R-C):
    - Prominence (R+C): Indicates the overall importance of the factor in the system.
    - Net Effect (R-C): Shows the net influence of the factor. A positive value indicates the factor is a net cause (influencer), while a negative value indicates it's a net effect (influenced).
8. Draw the Causal Diagram: A graph is plotted with (R + C) on the horizontal axis and (R-C) on the vertical axis. This diagram visually categorizes factors based on their influence and influenced status, helping identify key drivers and their relationships.

Advantages of DEMATEL
- Simple and easy to understand
- Handles complex interrelationships
- Provides a visual structural model
- Identifies key factors and their influence

Limitations of DEMATEL
- Relies on expert opinions, which can be subjective
- Potential for ambiguity in assessments
- Assumes linear relationships between factors

Variations of DEMATEL
- Fuzzy DEMATEL: Uses fuzzy logic to handle uncertainty and vagueness in assessments
- Grey DEMATEL: Used for analyzing grey data that has incomplete or uncertain information

Software for DEMATEL
- Microsoft Excel: For basic calculations and smaller matrices
- MATLAB: For more complex calculations and larger matrices
- Specialized multi-criteria decision-making software (e.g., Super Decisions, Expert Choice)

Key Takeaways: DEMATEL is a valuable tool for understanding complex systems by mapping cause-and-effect relationships and identifying key factors. The causal diagram provides a clear visualization of these relationships, aiding in decision-making and problem-solving.

2.6 DANP (DEMATEL-Based ANP)

DANP is a hybrid methodology that leverages the strengths of both DEMATEL and ANP. In this approach, DEMATEL is employed to ascertain the interdependence relationships among criteria and to construct the network structure for ANP. In other words, DEMATEL functions as a tool for building the ANP model. This integration enhances DANP's accuracy and efficiency compared to conventional ANP.

Advantages of Using DANP

- Considering Interdependencies: DANP is capable of accurately modeling complex relationships between criteria.
- Reducing Computational Complexity: By utilizing DEMATEL to determine the network structure, the need for extensive pairwise comparisons in ANP is reduced.
- Enhancing Accuracy and Validity of Results: The combination of DEMATEL and ANP leads to more accurate and valid results.

  In summary, DEMATEL serves as a tool for analyzing cause-and-effect relationships, ANP is a multi-criteria decision-making (MCDM) technique suitable for problems with interdependent relationships, and DANP is a composite of these two methods, inheriting the advantages of both. Table 2.10 provides a comparison of these three approaches.

2.7 Case Study

Determining the Intensity of Influence and Impact and Prioritizing the Factors Affecting Intellectual Capital Using a Hybrid Approach of Fuzzy DEMATEL and Fuzzy Network Process in Anzali Free Zone

Numerous models of intellectual capital have been proposed, each considering different dimensions for intellectual capital (Wee & Alton, 2013). Indicators are

Table 2.10 Comparison of three approaches DEMATEL, ANP, and DANP

Feature	DEMATEL	ANP	DANP
Primary objective	Analysis of cause-and-effect relationships	Multi-criteria decision-making with interdependencies	Multi-criteria decision-making with interdependencies (optimized by DEMATEL)
Problem structure	Causal diagram	Network	Network (constructed by DEMATEL)
Type of relationships	Cause-and-effect	Interdependency	Interdependency
Computational complexity	Relatively low	High	Medium
Accuracy and validity	Moderate	Good	Very good

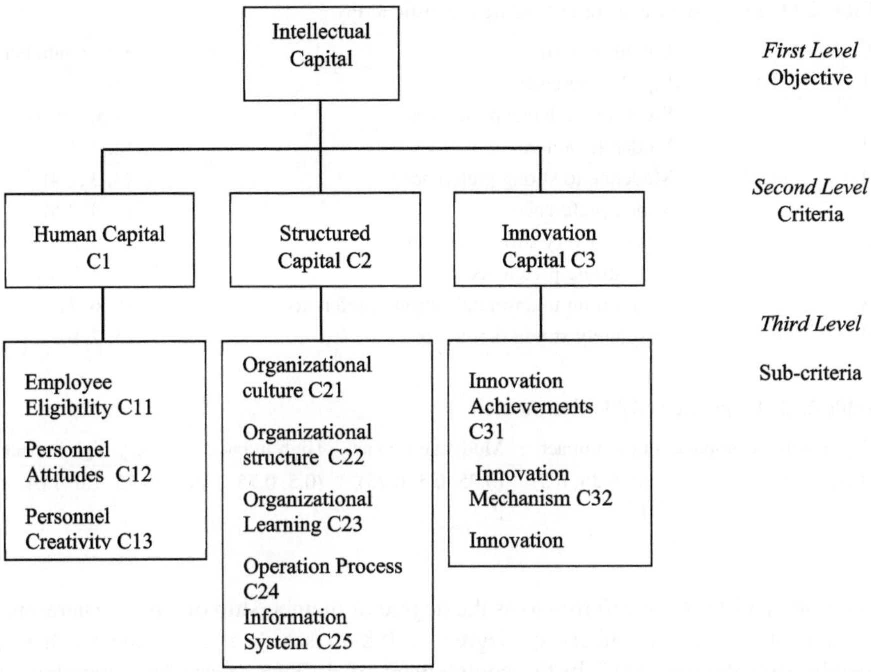

Fig. 2.1 Network analysis model

defined for each of these dimensions. The dimensions and indicators used in this research are based on the model developed by Chen et al., which includes three main criteria and 11 influential sub-criteria considered significant for the final model solution. The network analysis model is illustrated in Fig. 2.1.

2.7.1 Determining the Causal Relationships Among the Main Criteria of Intellectual Capital Using the Fuzzy DEMATEL Technique

Fuzzy Logic

Fuzzy logic is a mathematically grounded theory that generalizes the bivalent logic of set theory. In classical (crisp) set theory, an element x is either a member of set A or not. This implies that the characteristic function, or membership function, of x can take only two values: zero or one. A value of one indicates that x belongs to set A, while a value of zero indicates that x is outside the set. In contrast, fuzzy logic is a multi-valued logic system that allows for the expression of the degree to which elements such as x belong to a set through arbitrary membership functions taking continuous values in the range [0, 1]. The value of the characteristic or membership

Table 2.11 Fuzzy scale and corresponding linguistic terms

Code	Linguistic terms	Fuzzy number
1	Equal preference	(1, 1, 1)
2	Slight to moderate preference	(1.5, 1.5, 1)
3	Moderate preference	(2, 2, 1)
4	Moderate to strong preference	(3, 3.5, 4)
5	Strong preference	(3, 4, 4.5)
6	Strong to very strong preference	(3, 4.5, 5)
7	Very strong preference	(5, 5.5, 6)
8	Very strong to absolutely strong preference	(5, 6, 7)
9	Absolutely strong preference	(5, 7, 9)

Table 2.12 Linguistic variable classification

Absolutely no impact	Low impact	Moderate impact	High impact	Very high impact
(0, 0, 0.25)	(0, 0.25, 0.5)	(0.25, 0.5, 0.75)	(0.5, 0.75, 1.0)	(0.75, 1.0, 1.0)
0	1	2	3	4

function of x in set A is referred to as the degree of membership of x in A. Therefore, if element x has a membership degree of 0.8 in set A, it consequently has a membership degree of 0.2 in the complement of A. Thus, x can be a member of both sets (Dinara & Soufi, 2016).

Determining Fuzzy Numbers for Each Linguistic Variable

In this study, to determine the importance level of variables, a nine-point scale was used. The corresponding levels and fuzzy numbers are presented in Table 3.1, as proposed by Saati. Furthermore, to assess the impact of factors, a five-level scale was employed. These levels and their corresponding fuzzy numbers are presented in Table 3.2, based on the proposal by Lee (1999) (Safari & Soufi, 2014) (Tables 2.11 and 2.12).

Limitations of Traditional DEMATEL and the Necessity of Employing a Fuzzy Approach

Traditional DEMATEL is based on crisp numbers, meaning that precise and definite values are used to determine the influence of one factor on another. However, in many real-world problems, especially in multi-criteria decision-making (MCDM), expert opinions and available data are often associated with vagueness and uncertainty. For instance, when asking an expert, "What is the degree of influence of factor A on factor B?," their response might be in the form of linguistic terms such as "very high," "moderate," or "low." These linguistic terms are not precise or definite and cannot be easily converted into crisp numbers.

Traditional DEMATEL is incapable of processing such vague and uncertain data, which constitutes a significant limitation. To address this limitation, the fuzzy approach was integrated into DEMATEL, leading to the development of fuzzy DEMATEL.

Advantages of Using a Fuzzy Approach in DEMATEL

- **Modeling Vagueness and Uncertainty:** Fuzzy logic is capable of effectively modeling the vagueness and uncertainty inherent in expert opinions and data. By using fuzzy numbers (such as triangular or trapezoidal fuzzy numbers), linguistic terms can be represented mathematically and used in calculations.
- **Enhanced Flexibility:** Fuzzy DEMATEL offers greater flexibility compared to traditional DEMATEL and can cover a wider range of decision-making problems.
- **More Accurate and Realistic Results:** By considering vagueness and uncertainty in the data, the results obtained from fuzzy DEMATEL are usually more accurate and closer to reality.

Conditions for Using a Fuzzy Model (Generally and in DEMATEL)

In general, fuzzy models are used under the following conditions:

- **Presence of vagueness and uncertainty:** When the available data or opinions are associated with vagueness and uncertainty
- **Use of linguistic terms:** When linguistic terms are used to describe variables or relationships between them
- **System complexity:** When the system under study is complex and the relationships between its components are not precisely defined
- **Lack of precise and quantitative data:** When precise and quantitative data for modeling the system are not available

Specifically, in the DEMATEL method, fuzzy DEMATEL is used when:

- Expert opinions regarding the degree of influence of factors are expressed in linguistic terms.
- Vagueness and uncertainty exist in the data related to the relationships between factors.
- There is a need for a model that can effectively represent the complexity of relationships between factors, considering the existing vagueness.

In summary, fuzzy DEMATEL was developed to overcome the limitations of traditional DEMATEL in dealing with uncertainty and vagueness. The use of fuzzy logic in DEMATEL allows for more accurate and realistic modeling of the relationships between factors in situations where data and opinions are associated with vagueness (Safari & Soufi, 2014).

Phase 1: Constructing the Decision Matrix

To evaluate the criteria, the input of eight experts was considered. In these matrices, $X_{ij} = (L_{ij}, M_{ij}, U_{ij})$ represent triangular fuzzy numbers. $\widetilde{x}_{ij}(i = 1, 2, \ldots, n)$ is considered in the form of fuzzy numbers (0, 0, 0). To incorporate the opinions of all experts, the arithmetic mean is calculated according to below formula:

$$\widetilde{Z} = \frac{Z_1 + Z_2 + \ldots + Z_P}{P}.$$

In this formula, p represents the number of experts, and $\widetilde{x}^P, \widetilde{x}^2, \widetilde{x}^1$ represents the pairwise comparison matrices of expert 1, expert 2, and expert p, respectively. $\breve{Z}$ is a triangular fuzzy number in the form of $\widetilde{Z}_{ij} = \left(l'_{ij}, m'_{ij}, u'_{ij}\right)$:

$$\widetilde{x} \begin{vmatrix} \widetilde{x}_{11} & \widetilde{x}_{12} & \cdots & \widetilde{x}_{1n} \\ \widetilde{x}_{21} & \widetilde{x}_{22} & \cdots & \widetilde{x}_{2n} \\ \vdots & \vdots & \ddots & \vdots \\ \widetilde{x}_{n1} & \widetilde{x}_{n1} & \cdots & \widetilde{x}_{nn} \end{vmatrix}.$$

Normalization of the obtained matrix was performed employing the formulas below:

$$\widetilde{x}_{ij} = \frac{\widetilde{Z}_{ij}}{r} = \left(\frac{l_{ij}}{r}, \frac{m_{ij}}{r}, \frac{u_{ij}}{r}\right)$$

$$r = \max\ 1 \le i \le u\left(\sum\nolimits_{j=1}{}^{n} u_{ij}\right).$$

Upon calculating the previously mentioned matrices, the fuzzy total relation matrix is derived using the following formulas:

$$T = \lim_{k \to +\infty} \left(\widetilde{H}^1 \oplus \widetilde{H}^2 \oplus \ldots \oplus \widetilde{H}^k\right)$$

where each element is a fuzzy number in the form of $\widetilde{t}_{ij} = \left(l^t_{ij}, m^t_{ij}, u^t_{ij}\right)$ and is estimated in the following manner:

$$\begin{aligned} \left[l^t_{ij}\right] &= H_l \times (I - H_l)^{-1} \\ \left[m^t_{ij}\right] &= H_m \times (I - H_m)^{-1}. \\ \left[u^t_{ij}\right] &= H_u \times (I - H_u)^{-1} \end{aligned}$$

In these formulas, I is the identity matrix, and H_l, H_m, and H_u are each $n \times n$ matrices whose entries are formed by the lower, middle, and upper values, respectively, of the triangular fuzzy numbers of matrix H. Specifically, if $H_{ij} = (l_{ij}, m_{ij}, u_{ij})$ is a triangular fuzzy number in matrix H, then $(H_l)_{ij} = l_{ij}$, $(H_m)_{ij} = m_{ij}$, and $(H_u)_{ij} = u_{ij}$.

The following step involves calculating the row and column sums of the matrix $\widetilde{t}$. These sums are calculated using the following formulas:

$$\widetilde{D} = \left(\widetilde{D}_i\right)_{n\times 1} = \left[\sum_{j=1}^{n} \widetilde{T}_{ij}\right]_{n\times 1}$$

$$\widetilde{R} = \left(\widetilde{R}_i\right)_{1\times n} = \left[\sum_{j=1}^{n} \widetilde{T}_{ij}\right]_{1\times n}$$

where R and D are respectively $1 \times n$ and $n \times 1$ matrices. The next step reveals the importance of the indices $\left(\widetilde{D}_i + \widetilde{R}_i\right)$ and $\left(\widetilde{D}_i - \widetilde{R}_i\right)$ and the relationships between them. If $\widetilde{D}_i - \widetilde{R}_i > 0$, the corresponding criterion is influential (or has an influence), and if $\widetilde{D}_i - \widetilde{R}_i < 0$, the corresponding criterion is influenced (or is influenced by other criteria).

In the subsequent step, the fuzzy numbers $\widetilde{D}_i + \widetilde{R}_i$ and $\widetilde{D}_i - \widetilde{R}_i$ obtained from the previous step are defuzzified according to:

$$B = \frac{a_1 + a_2 + 2 \times a_2}{4},$$

where B is the defuzzified value of the fuzzy number $\widetilde{A} = (a_1, a_2, a_3)$.

Phase 2: Steps for Obtaining Component Weights Using the Fuzzy Analytical Network Process (FANP) Technique

The methodology for deriving component weights from the supermatrix comprises the following stages:

Stage 1: Aggregation of expert judgments is performed by employing the geometric mean of the respondents' pairwise comparisons.

Stage 2: Eigenvector determination: The logarithmic least squares method is utilized to determine the eigenvector of each aggregated pairwise comparison matrix, as defined by following relations:

$$W_k^s = \frac{\left(\prod_{j=1}^{n} a_{kj}^s\right)^{\frac{1}{n}}}{\sum_{j=1}^{n}\left(\prod_{j=1}^{n} a_{kj}^m\right)} \quad s \in \{l, m, u\}$$

$$\widetilde{W}_k = \left(w_k^l, w_k^m, w_k^u\right) \; k = 1, 2, \ldots, n.$$

The third stage consists of the formation of eigenvector matrices (W_{ij}). These matrices encompass the eigenvectors resulting from the pairwise comparisons conducted in stage 2. It is noteworthy that any deviation from (1, 1, 1) on the main diagonal of the inner dependence eigenvector matrix signifies that the respective column has been normalized. The normalization procedure entails dividing each fuzzy number within the column by the summation of the model values (middle values) of the fuzzy numbers within that same column.

Phase 3: Combining ANP and DEMATEL for Ranking Factors and Determining the Final Weights of Criteria and Sub-Criteria with Respect to the Objective

In this step, the eigenvectors obtained from fuzzy ANP are multiplied by the fuzzy total-relation matrix obtained from DEMATEL. To perform this matrix multiplication, the total-relation matrix of the main criteria derived from fuzzy DEMATEL is first normalized. Then, the eigenvector matrix (weights) of the criteria with respect to the objective is multiplied by the normalized total-relation matrix of the main criteria to obtain the final weights of the main criteria with respect to the objective.

To determine the final weights of the sub-criteria, the normalized total-relation matrix of the sub-criteria must first be obtained. To determine the final weights of the indicators with respect to the objective, their weights (eigenvectors) should be multiplied by the normalized total-relation matrix of the sub-criteria, and the result should be multiplied by the final weights of the main criteria with respect to the objective. In this research, the method of Gougousis and Boucher has been used to calculate consistency.

2.7.2 *Results of Fuzzy DEMATEL Technique*

The analysis of internal relationships among criteria using the F-DEMATEL method is as follows:

2.7.2.1 Determining Causal Relationships Among the Main Criteria of Intellectual Capital

Table 2.13 presents the fuzzy average of expert opinions regarding the impact of human capital, structural capital, and customer capital. To incorporate the perspectives of all experts, the arithmetic mean is calculated.

Table 2.14 presents the normalized matrix of the main factors of intellectual capital.

Table 2.15 presents the total relation matrix (t).

The next step is to obtain the row and column sums of the T~ matrix. Subsequently, the importance of the criteria $\left(\widetilde{D}_i + \widetilde{R}_i\right)$ and the relationships among the attributes $\left(\widetilde{D}_i + \widetilde{R}_i\right)$ are determined. If $\widetilde{D}_i - \widetilde{R}_i > 0$, the corresponding criterion is

Table 2.13 Average fuzzy expert opinions on the impact of main factors

Average opinion of all experts	Human capital	Structural capital	Innovation capital
Human capital	(0, 0, 0)	(0.40, 0.56, 0.87)	(0.46, 0.68, 0.81)
Structural capital	(0.12, 0.28, 0.53)	(0, 0, 0)	(0.25, 0.43, 0.62)
Innovation capital	(0.15, 0.37, 0.62)	(0.40, 0.65, 0.81)	(0, 0, 0)

Table 2.14 Normalized matrix of the three factors

Normalized matrix	Human capital	Structural capital	Innovation capital
Human capital	(0, 0, 0)	(0.24, 0.38, 0.51)	(0.27, 0.40, 0.48)
Structural capital	(0.07, 0.16, 0.31)	(0, 0, 0)	(0.14, 0.25, 0.37)
Innovation capital	(0.09, 0.22, 0.37)	(0.24, 0.38, 0.48)	(0, 0, 0)

Table 2.15 Total relation matrix of main criteria

Total relation matrix	Human capital	Structural capital	Innovation capital
Human capital	(0.34, 0.73, 2.00)	(0.33, 0.78, 2.23)	(0.05, 0.29, 1.44)
Structural capital	(0.18, 0.47, 1.55)	(0.06, 0.30, 1.44)	(0.09, 0.32, 1.34)
Innovation capital	(0.07, 0.34, 1.49)	(0.28, 0.68, 2.00)	(0.12, 0.41, 1.55)

Table 2.16 Importance and influence of criteria (fuzzy numbers)

Criteria	$\widetilde{D}_i + \widetilde{R}_i$	$\widetilde{D}_i - \widetilde{R}_i$
Human capital	(1.01, 2.84, 10.02)	(−4.5, 0.10, 4.44)
Structural capital	(1.04, 2.88, 10.02)	(−5.33, 0.67, 3.56)
Innovation capital	(1.09, 2.99, 10.09)	(−3.66, 0.78, 5.40)

Table 2.17 Importance and influence of criteria (crisp numbers)

Criteria	$\left(\widetilde{D}_i + \widetilde{R}_i\right)^{def}$	$\left(\widetilde{D}_i - \widetilde{R}_i\right)^{def}$
Human capital	4.18	0.84
Structural capital	4.2	−0.75
Innovation capital	4.29	−0.08

influential (influencing), and if $\widetilde{D}_i - \widetilde{R}_i < 0$, the corresponding criterion is influenced (receiving). Table 2.16 presents $\widetilde{D}_i + \widetilde{R}_i$ and $\widetilde{D}_i - \widetilde{R}_i$.

Table 2.17 presents the defuzzified values of Table 2.16.

Based on Table 2.17, as observed, the innovation capital criterion exhibits greater interaction and importance compared to other dimensions within the dynamics of influencing and being influenced by criteria. This intensity of effects simplifies the input for the analytical network process (ANP) technique, enabling us to achieve more straightforward and accurate results compared to complex questionnaires.

2.7.2.2 Determining Causal Relationships Among Sub-Criteria of Main Factors

Based on the relationships established in the preceding steps, calculations for the normalized pairwise comparison matrices, total-relation matrices, and importance and influence (fuzzy and crisp) values for the sub-criteria of the main factors have been performed, as detailed in Tables 2.18, 2.19, 2.20, and 2.21.

Table 2.18 Normalized matrix of human capital sub-criteria

	Employee creativity	Employee attitude	Employee competence
Employee competence	(0, 0, 0)	(0.27, 0.40, 0.50)	(0.26, 0.39, 0.49)
Employee attitude	(0.03, 0.09, 0.22)	(0, 0, 0)	(0.14, 0.24, 0.34)
Employee creativity	(0.13, 0.22, 0.36)	(0.31, 0.44, 0.50)	(0, 0, 0)

Table 2.19 Matrix of total relationships of human capital sub-criteria

	Staff creativity	Staff attitude	Staff competency
Staff competency	(0.38, 0.81, 1.87)	(0.35, 0.55, 0.67)	(0.10, 0.35, 1.21)
Staff attitude	(0.35, 0.71, 1.69)	(0.07, 0.28, 1.06)	(0.20, 0.47, 1.32)
Staff creativity	(0.06, 0.25, 1.06)	(0.08, 0.28, 1.03)	(0.15, 0.35, 1.10)

Table 2.20 Importance and impact of human capital sub-criteria (fuzzy numbers)

Criteria	$\widetilde{D}_i + \widetilde{R}_i$	$\widetilde{D}_i - \widetilde{R}_i$
Staff competency	(1.30, 3.03, 8.28)	(−2.78, 0.65, 4.15)
Staff attitude	(1.14, 2.71, 7.73)	(−3.01, 0.23, 3.56)
Staff creativity	(1.10, 2.67, 7.89)	(−4.34, 0.88, 2.39)

Table 2.21 Importance and impact of human capital sub-criteria (crisp numbers)

Criteria	$\left(\widetilde{D}_i + \widetilde{R}_i\right)^{def}$	$\left(\widetilde{D}_i + \widetilde{R}_i\right)^{def}$
Staff competency	3.9148	0.6763
Staff attitude	3.5796	0.2538
Staff creativity	3.5757	−0.093

2.7.2.2.1 Determining the Causal Relationships Between Human Capital Sub-Criteria

Based on Table 2.21, the staff competency sub-criterion, within the context of sub-criteria influence and being influenced, holds greater importance compared to other human capital sub-criteria. In summary, the following results can be extracted:

- Staff competency is the most significant factor (highest $D + R$).
- Staff competency and staff attitude primarily play an influencing role (positive D-R).
- Staff creativity predominantly has an influenced role (negative $D - R$) (Tables 2.22, 2.23, and 2.24).

2.7.2.2.2 Determining the Causal Relationships Between Structural Capital Sub-Criteria

As shown in Table 2.25, the operational process sub-criterion exhibits the highest value among all sub-criteria. This suggests its greater importance within the influence flow of structural capital sub-criteria.

Table 2.22 Normalized matrix of structural capital sub-criteria

	Culture	Structure	Learning	Operational process	Information system
Culture	(0, 0, 0)	(0.099, 0.16, 0.24)	(0.30, 0.08, 0.16)	(0.0.09, 0.16, 0.24)	(0.13, 0.21, 0.27)
Structure	(0.07, 0.12, 0.19)	(0, 0, 0)	(0.01, 0.07, 0.15)	(0.08, 0.14, 0.21)	(0.03, 0.09, 0.17)
Learning	(0.10, 0.18, 0.25)	(0.009, 0.04, 0.12)	(0, 0, 0)	(0.11, 0.19, 0.25)	(0.10, 0.17, 0.23)
Operational process	(0.03, 0.09,0.17)	(0.11, 0.19, 0.26)	(0.12, 0.20, 0.27)	(0, 0, 0)	(0.15, 0.23, 0.28)
Information system	(0.02, 0.08, 0.16)	(0.009, 0.04, 0.12)	(0.14, 0.22, 0.27)	(0.03, 0.10, 0.18)	(0, 0, 0)

Table 2.23 Total relations matrix of structural capital sub-criteria

	Culture	Structure	Learning	Operational process	Information system
Culture	(0.02, 0.15, 1.08)	(0.11, 0.29, 1.24)	(0.08, 0.28, 1.34)	(0.12, 0.33, 1.40)	(0.17, 0.40, 1.50)
Structure	(0.09, 0.23, 1.07)	(0.02, 0.11, 0.87)	(0.04, 0.22, 1.13)	(0.10, 0.27, 1.18)	(0.07, 0.26, 1.23)
Learning	(0.12, 0.31, 1.25)	(0.04, 0.02, 1.12)	(0.04, 0.20, 1.15)	(0.14, 0.36, 1.36)	(0.15, 0.38, 1.43)
Operational process	(0.07, 0.29, 0.17)	(0.13, 0.32, 1.31)	(0.16, 0.04, 1.48)	(0.04, 0.22, 1.27)	(0.19, 0.45, 1.58)
Information system	(0.005, 0.02, 1.06)	(0.02, 0, 1.61)	(0, 1.63, 1.33)	(0.06, 0.25, 1.18)	(0.03, 0.18, 1.09)

Table 2.24 Importance of structural capital sub-criteria impact (fuzzy numbers)

Criteria	$\widetilde{D}_i + \widetilde{R}_i$	$\widetilde{D}_i - \widetilde{R}_i$
Organizational culture	(0.91, 2.69, 12.35)	(−5.25, −0.27, −12.35)
Organizational structure	(0.68, 2.21, 11.70)	(−5.21, −0.01, −5.17)
Organizational learning	(1.02, 2.94, 12.68)	(−5.83, 0.0, 5.82)
Operational process	(1.11, 3.13, 13.36)	(−5.79, 0.22, 6.45)
Information system	(0.98, 2.86, 12.44)	(−6.5, 0.5, 4.9)

Table 2.25 Impact importance of structural capital sub-criteria (crisp numbers)

Criteria	$\left(\widetilde{D}_i + \widetilde{R}_i\right)^{def}$	$\left(\widetilde{D}_i - \widetilde{R}_i\right)^{def}$
Organizational culture	4.66	0.37
Organizational structure	4.05	−0.003
Organizational learning	4.90	−0.002
Operational process	5.18	0.27
Information system	4.78	−0.65

Table 2.26 Normalized matrix of innovation capital sub-criteria

Criteria	Innovation culture	Innovation mechanism	Innovation outcomes
Innovation culture	(0, 0, 0)	(0.07, 0.19, 0.35)	(0.23, 0.39, 0.50)
Innovation mechanism	(0.09, 0.25, 0.41)	(0, 0, 0)	(0.31, 0.47, 0.58)
Innovation outcomes	(0.07, 0.21, 0.37)	(0.29, 0.45, 0.56)	(0, 0, 0)

Table 2.27 Total relations matrix of innovation capital sub-criteria

Criteria	Innovation culture	Innovation mechanism	Innovation outcomes
Innovation culture	(0.07, 0.44, 4.33)	(0.16, 0.60, 4.47)	(0.26, 0.71, 4.56)
Innovation mechanism	(0.35, 0.85, 5.16)	(0.08, 0.49, 4.63)	(0.18, 0.71, 4.95)
Innovation outcomes	(0.18, 0.69, 4.92)	(0.33, 0.80, 4.87)	(0.07, 0.47, 4.51)

Table 2.28 Importance and impact of innovation capital sub-criteria (fuzzy numbers)

Criteria	$\widetilde{D}_i + \widetilde{R}_i$	$\widetilde{D}_i - \widetilde{R}_i$
Innovation culture	(1.11, 3.76, 27.78)	(−13.91, −12.75, 22)
Innovation mechanism	(1.20, 3.96, 28.74)	(−13.35, 0.16,14.7)
Innovation outcomes	(1.13, 3.89, 28.34)	(−13.43, 0.06, 13.77)

Table 2.29 Importance and impact of innovation capital sub-criteria (crisp numbers)

Criteria	$\left(\widetilde{D}_i + \widetilde{R}_i\right)^{def}$	$\left(\widetilde{D}_i - \widetilde{R}_i\right)^{def}$
Innovation culture	9.1	−0.4
Innovation mechanism	9.46	0.28
Innovation outcomes	9.31	0.11

Table 2.25 presents the impact importance of structural capital sub-criteria, using crisp numbers. The $\left(\widetilde{D}_i + \widetilde{R}_i\right)^{def}$ column likely represents the initial weight or importance of each sub-criterion, where negative values indicate a potentially adverse or inverse effect. The $\left(\widetilde{D}_i - \widetilde{R}_i\right)^{def}$ column, on the other hand, reflects the ultimate influence of each sub-criterion on the overall structural capital. As observed in the table, the operational process sub-criterion exhibits the highest impact importance (5.18) among all sub-criteria. This signifies that the operational process plays the most significant role in influencing the overall structural capital, despite having a moderate initial value (0.27) (Table 2.26, 2.27, 2.28, 2.29, 2.30, 2.31, 2.32, 2.33, 2.34, 2.35, and 2.36).

2.7.2.2.3 Determining Causal Relationships Among Sub-Criteria of Innovation Capital

Based on this, it can be concluded that:

Innovation culture: This criterion exhibits a relatively wide range of variation (−0.40 to 9.10). The negative value in the lower bound may indicate a negative impact

Table 2.30 Mean of pairwise comparisons with respect to intellectual capital

Criteria	Human capital	Structural capital	Customer capital	Eigenvector
Human capital	(1, 1, 1)	(0.82, 0.82, 1)	(1.73, 2.14, 2.29)	(0.35, 0.38, 0.41)
Structural capital	(1, 1.20, 1.20)	(1, 1, 1)	(1.51, 1.93, 2.03)	(0.36, 0.42, 0.42)
Customer capital	(0.43, 0.46, 0.75)	(0.49, 0.51, 0.66)	(1, 1, 1)	(0.19, 0.19, 0.23)
Consistency ratio (CR)	$\mathbf{CR^g = 0.016}$	$\mathbf{CR^m = 0.01}$	consistent	

Table 2.31 Mean of pairwise comparisons with respect to intellectual capital

Criteria	Employee competence	Employee attitude	Employee creativity	Eigenvector
Employee competence	(1, 1, 1)	(0.82, 0.82, 1)	(1, 1.41, 1.41)	(0.28, 0.31, 0.33)
Employee attitude	(1.31, 1.56, 1.62)	(1, 1, 1)	(1, 1, 1.1)	(0.36, 0.39, 0.40)
Employee creativity	(0.70, 0.70, 1)	(0.90, 0.90, 1)	(1, 1, 1)	(0.28, 0.28, 0.33)
Consistency ratio (CR)	$\mathbf{CR^g = 0.078}$	$\mathbf{CR^m = 0.55}$	consistent	

or the absence of achievement under certain conditions. The high value of 9.10 in the upper bound suggests that this criterion can have a very significant impact. This wide range demonstrates that the impact of innovation achievements is highly variable and depends on various factors.

Innovation mechanism: This criterion has a lower bound of 0.28 and an upper bound of 9.46. This indicates that the innovation mechanism always has a positive impact (even if small) and can be highly influential. The relatively wide range of this criterion also indicates its susceptibility to various factors.

Innovation outcomes: This criterion has a lower bound of 0.11 and an upper bound of 9.31. Similar to the innovation mechanism, this criterion also always has a positive impact and can be highly influential.

2.7.3 Determining the Eigenvector of the Main Factors Relative to the Goal and the Sub-Factors Relative to the Main Factors by Employing Fuzzy ANP

Following the acquisition of DEMATEL outputs, there are several essential steps to obtain the eigenvector after establishing the ANP network:

Table 2.32 Mean of pairwise comparisons with respect to structural capital

Criteria	Organizational culture	Organizational structure	Learning	Operational process	Information system	Eigenvector
Organizational culture	1.14, 1.448, 1.51	1.14, 1.6, 1.70	0.73, 0.74, 0.87	1, 1.16, 1.16	1, 1, 1	0.19, 0.22, 0.23
Organizational structure	1, 1, 1	1.14, 1.41, 1.4	1.14, 1.74, 1.76	1, 1.14, 1.14	0.66, 0.67, 0.87	0.19, 0.22, 0.23
Learning	0.69, 0.70, 0.87	1, 1, 1	0.53, 0.54, 0.87	0.53, 0.60, 0.8	0.56, 0.56, 0.87	0.12, 0.13, 0.17
Operational process	0.56, 0.56, 0.87	1.14, 1.83, 1.86	1, 1, 1	1, 1.20, 1.20	1.14, 1.34, 1.36	0.18, 0.21, 0.23
Information system	0.87, 0.87, 1	1.86, 1.65, 1.20	1, 0.82, 0.82	1, 1, 1	0.85, 0.85, 1	0.22, 0.19, 0.18
Consistency ratio (CR)	$\mathbf{CR^g = 0.39}$	$\mathbf{CR^m = 0.29}$ consistent				

Table 2.33 Mean of pairwise comparisons with respect to innovation capital

Criteria	Innovation achievements	Innovation mechanism	Innovation culture	Eigenvector
Innovation achievements	(1, 1, 1)	(1, 1.73, 1.73)	(0.82, 0.82, 1)	(0.29, 0.35, 0.38)
Innovation mechanism	(0.57, 0.57, 1)	(1, 1, 1)	(0.46, 0.48, 0.76)	(0.20, 0.20, 0.29)
Innovation culture	(1, 1.20, 1.20)	(1.31, 2.06, 2.13)	(1, 1, 1)	(0.34, 0.43, 0.43)
Consistency ratio (CR)	$\mathbf{CR^g = 0.007}$	$\mathbf{CR^m = 0}$ consistent		

Table 2.34 Eigenvector matrix of level 2 relative to level 1

Intellectual capital	
Human capital	(0.35, 0.38, 0.41)
Structural capital	(0.36, 0.42, 0.42)
Innovation capital	(0.19, 0.19, 0.23)

Table 2.35 Eigenvector matrix of level 3 relative to level 2

Criteria	Human capital	Structural capital	Innovation capital
Employee competence	(0.28, 0.31, 0.33)	(0, 0, 0)	(0, 0, 0)
Employee attitude	(0.36, 0.39, 0.40)	(0, 0, 0)	(0, 0, 0)
Employee creativity	(0.28, 0.28, 0.33)	(0, 0, 0)	(0, 0, 0)
Organizational culture	(0, 0, 0)	(0.19, 0.22, 0.23)	(0, 0, 0)
Organizational structure	(0, 0, 0)	(0.19, 0.22, 0.23)	(0, 0, 0)
Organizational learning	(0, 0, 0)	(0.12, 0.13, 0.17)	(0, 0, 0)
Operational processes	(0, 0, 0)	(0.18, 0.21, 0.23)	(0, 0, 0)
Information systems	(0, 0, 0)	(0.18, 0.19, 0.22)	(0, 0, 0)
Innovation achievements	(0, 0, 0)	(0, 0, 0)	(0.29, 0.35, 0.38)
Innovation mechanisms	(0, 0, 0)	(0, 0, 0)	(0.20, 0.20, 0.29)
Innovation culture	(0, 0, 0)	(0, 0, 0)	(0.34, 0.43, 0.43)

Table 2.36 Normalized total relation matrix of main factors

Criteria	Human capital	Structural capital	Innovation capital
Human capital	(0.01, 0.28, 5.27)	(0.09, 0.44, 3.22)	(0.06, 0.47, 3.31)
Structural capital	(0.02, 0.31, 4.90)	(0.018, 0.18, 2.08)	(0.03, 0.30, 2.56)
Innovation capital	(0.02, 0.40, 2.56)	(0.07, 0.38, 2.89)	(0.01, 0.22, 2.46)
	(0.27, 1.02, 4.34)	(0.69, 1.78, 3.68)	(0.60, 1.54, 5.04)

1. Expert selection: A group of experts is assembled to provide informed judgments on the pairwise comparisons. These experts should possess a deep understanding of the decision-making subject matter and be capable of assessing the relative importance of various factors involved.

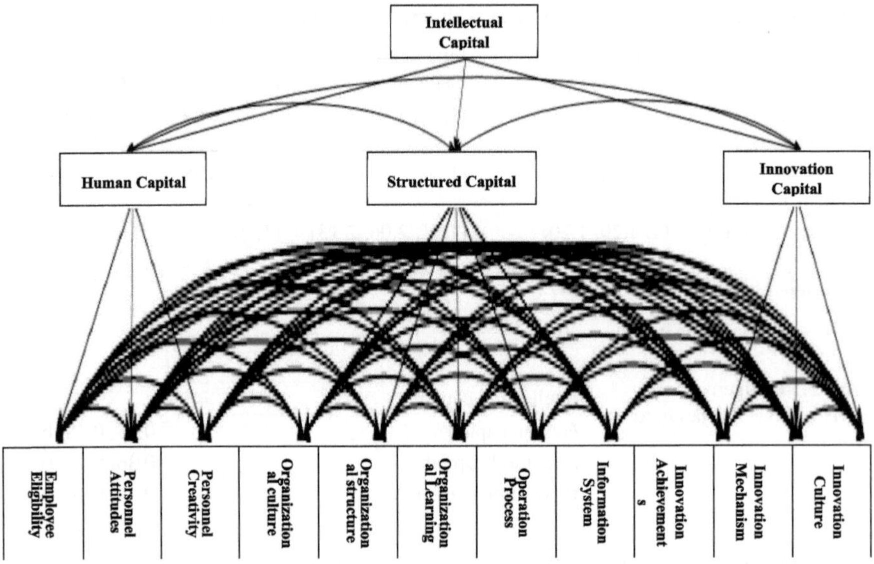

Fig. 2.2 Research network diagram

2. Pairwise comparisons: The experts are then tasked with conducting pairwise comparisons. To facilitate this, pairwise comparison matrices are constructed for both the main and sub-factors. The experts are then asked to evaluate the relative importance of each pair of factors. For instance, if we have two factors, A and B, the experts would determine whether A is more important than B, B is more important than A, or if both factors hold equal importance. These judgments are typically captured using a Likert scale or another suitable preference scale.
3. Aggregation of pairwise comparison matrices: The individual pairwise comparison matrices need to be aggregated to reflect a consensus view. This involves combining the judgments provided by all experts for each pair of factors. Common methods for aggregation include the arithmetic mean, the geometric mean, or other relevant techniques.
4. Eigenvector calculation: The aggregated pairwise comparison matrices are then used to compute the eigenvector. The eigenvector provides the relative weights of each factor, indicating their respective importance in the decision-making process. This computation can be performed using various software tools such as Expert Choice or Super Decisions.
5. Determining the eigenvector: Finally, the eigenvector of the main factors with respect to the overall goal and the eigenvectors of the sub-factors with respect to their respective main factors are determined. This is achieved by multiplying the appropriate eigenvector matrices (Fig 2.2).

2.7.3.1 Steps for Obtaining Component Weights Using Fuzzy Network Analysis Based on Supermatrix

To calculate the component weights, the following steps are taken:

Step 1: Aggregating Expert Opinions

To consolidate expert opinions, the geometric mean of respondents' pairwise comparisons is calculated.

Step 2: Calculating the Eigenvector

To calculate the eigenvector for each aggregated pairwise comparison table, the logarithmic least squares method is employed:

$$W_k^s = \frac{\left(\prod_{j=1}^{n} a_{kj}^s\right)^{\frac{1}{n}}}{\sum_{j=1}^{n}\left(\prod_{j=1}^{n} a_{kj}^m\right)} \quad s \in \{l, m, u\}$$

$$\widetilde{W}_k^s = \left(w_k^l, w_k^m, w_k^u\right) \; k = 1, 2, \ldots, n.$$

In the last column of the tables below, the eigenvector and all pairwise comparison matrices that include the vectors and their consistency ratio are shown.

In general, these matrices can be divided into two categories:

1. Matrices that encompass eigenvectors representing vertical (orthogonal) relationships between surfaces. If there is no vertical relationship between two components, the value (0, 0, 0) is placed at the intersection of those two components in the matrix. In other entries, based on the orthogonal relationship of the components, the values of the eigenvectors obtained from the second stage are placed.
2. Matrices that include eigenvectors representing horizontal (intra-surface) relationships. These matrices are square, and their main diagonal is (1, 1, 1). If there is no horizontal relationship between two components, the value (0, 0, 0) is placed at the intersection of those two components in the matrix. In other entries, based on the horizontal relationship of the components, the values of the eigenvectors obtained from the second stage are placed. If one or more entries in the main diagonal of the intra-surface eigenvector matrix are not (1, 1, 1), it is because normalization has been applied to that column. Normalization is performed by dividing all fuzzy numbers in that column by the sum of the median values of the fuzzy numbers in that column.

   The following tables present the eigenvector matrices.

At this stage, the eigenvectors obtained from the fuzzy analytic network process (fuzzy ANP) are multiplied by the total relation matrix derived from the fuzzy DEMATEL method. To perform this multiplication, the total relation matrix of the main factors obtained from fuzzy DEMATEL is first normalized. This normalization ensures the scaling consistency and prepares the matrix for the matrix multiplication operation. After normalization, the normalized matrix is multiplied by the

Table 2.37 Final weights of main factors relative to the goal

Main factor	Fuzzy weight	Crisp weight	Rank
Human capital	(1.90, 0.15, 8.001)	2.550	1
Structural capital	(0.85, 0.11, 6.630)	1.925	2
Innovation capital	(2.14, 0.06, 4.11)	1.592	3

Table 2.38 Normalized total relation matrix of human capital sub-criteria

Sub-criteria	Employee competence	Employee attitude	Employee creativity
Employee competence	(0.28, 0.29, 2.63)	(0.09, 0.54, 3.02)	(0.83, 0.045, 2.33)
Employee attitude	(0.056, 0.40, 2.78)	(0.02, 0.22, 2.06)	(0.07, 0.40, 2.1)
Employee creativity	(0.04, 0.30, 2.39)	(0.02, 0.02, 2.01)	(0.013, 0.014, 1.32)
	(0.46, 1.1, 3.63)	(0.51, 1.24, 3.65)	(0.80, 1.78, 4.64)

Table 2.39 Final weight of human capital sub-criteria relative to the objective

Sub-criteria	Fuzzy weight	Crisp weight	Rank
Employee competence	(0.302, 0.137, 0.016)	0.148	3
Employee attitude	(1.95, 0.13, 2.66)	1.217	1
Employee creativity	(1.30, 0.063, 1.057)	0.621	2

eigenvectors from fuzzy ANP, and the final results are calculated. This process aids in analyzing and prioritizing factors within complex systems.

Subsequently, the eigenvector matrix of the criteria level relative to the goal level is multiplied by the normalized total relation matrix of the main factors. After multiplying the aforementioned matrix in Table 2.34, Table 2.37 is obtained, which represents the final weights of the main factors relative to the goal.

To determine the final weights of the sub-criteria, we must first obtain the normalized total relation matrix of the sub-criteria (Table 2.38).

To determine the final weights of the indicators relative to the goal, their weights must be multiplied by the normalized total relation matrix of the sub-criteria, and the resulting values must then be multiplied by the final weights of the main factors relative to the goal. After multiplying the aforementioned matrix in Table 2.35, Table 2.39 is obtained, which presents the final weights of the human capital sub-criteria relative to the goal.

To determine the final weights of the sub-criteria of structural capital, we must first obtain the normalized total relation matrix of the sub-criteria (Table 2.40).

After multiplying the above matrix by Table 2.35, Table 2.41 is obtained, which represents the final weights of the sub-criteria of structural capital relative to the goal.

Similarly, to determine the final weights of the sub-criteria of innovation capital, we must first obtain the normalized total relation matrix of the sub-criteria (Table 2.42).

After multiplying the above matrix by Table 2.35, Table 2.43 is obtained, which represents the final weights of the sub-criteria of innovation capital relative to the goal.

Table 2.40 Normalized total relation matrix of the sub-criteria of structural capital

Sub-criteria	Organizational culture	Organizational structure	Organizational learning	Operational process	Information system
Organizational culture	(0.13, 2.8, 4.9)	(0.02, 0.02, 2.2)	(0.01, 0.19, 2.61)	(0.01, 0.23, 2.82)	(0.02, 0.02, 2.3)
Organizational structure	(0.019, 0.15, 2.8)	(0.01, 0.01, 4.1)	(0.015, 0.015, 7.62)	(0.01, 0.19, 2.3)	(0.01, 0.015, 1.9)
Organizational learning	(0.02, 0.26, 0.85)	(0.18, 0.18, 7.40)	(0.13, 0.13, 7.4)	(0.02, 0.02, 2.75)	(0.02, 0.22, 2.2)
Operational process	(0.01, 0.22, 3.4)	(0.02, 0.29, 2.39)	(0.02, 0.27, 2.89)	(0.15, 0.15, 7.20)	(0.020, 0.026, 2.47)
Information system	(0.17, 0.17, 9.12)	(0.14, 0.14, 4.45)	(0.02, 0.23, 2.41)	(0.01, 0.17, 2.38)	(0.10, 0.10, 5.30)
	(0.37, 1.20, 5.77)	(0.54, 1.1, 5.5)	(0.51, 1.47, 6.35)	(0.49, 1.45, 6.41)	(0.63, 1.69, 6.85)

Table 2.41 Final weights of the sub-criteria of structural capital relative to the goal

Sub-criteria	Fuzzy weight	Crisp weight	Rank
Organizational culture	(0.19, 0.22, 0.23)	0.215	2
Organizational structure	(0.19, 0.22, 0.23)	0.215	2
Organizational learning	(0.12, 0.13, 0.017)	0.099	4
Operational process	(0.18, 0.21, 0.23)	0.217	1
Information system	(0.18, 0.19, 0.22)	0.195	3

Table 2.42 Normalized total relation matrix of the sub-criteria of innovation capital

Sub-criteria	Innovation achievements	Innovation mechanism	Innovation culture
Innovation culture	(0.01, 0.34, 7.99)	(0.02, 0.42, 8.43)	(0.24, 5.59, 8.44)
Innovation mechanism	(0.02, 0.42, 8.38)	(0.25, 5.9, 8.02)	(0.01, 0.37, 9.26)
Innovation achievements	(0.019, 0.37, 8.5)	(0.011, 0.31, 7.74)	(0.22, 5, 7.036)
	(0.61, 1.99, 14.41)	(0.57, 1.90, 13.98)	(0.53, 1.91, 13.73)

Table 2.43 Final weights of the sub-criteria of innovation capital relative to the goal

Sub-criterion	Fuzzy weight	Crisp weight	Rank
Innovation achievements	(0.109, 0.203, 5.34)	1.440	1
Innovation culture	(0.146, 0.265, 0.650)	0.302	2
Innovation mechanism	(0.073, 0.167, 0.447)	0.190	3

2.8 The Best-Worst Method (BWM)

The best-worst method (BWM) was introduced by Jafar Rezaei in 2015. This method was proposed as a novel approach in multi-criteria decision-making (MCDM) and quickly gained attention due to its simplicity, accuracy, and reduced need for extensive pairwise comparisons. BWM is based on the idea that instead of comparing all criteria or alternatives pairwise, only the best (most important) and worst (least important) criterion or alternative is identified, and then pairwise comparisons are made solely between these two and the other criteria. This approach significantly reduces the computational burden and enhances the accuracy of the results. The more straightforward structure of BWM, as opposed to AHP, leads to greater consistency in the obtained results.

Based on this, pairwise comparisons can be classified into two categories:

1. Base comparisons
2. Secondary comparisons

Definition 1: A comparison a_{ij} is a **base comparison** if i is the best element **or** j is the worst element.

Definition 2: A comparison a_{ij} is a **secondary comparison** if **neither** *i* **nor** *j* is the best or worst element **and** $a_{ij} > 1$.

2.8.1 Case Study

2.8.1.1 Identification and Prioritization of Entrepreneurial Opportunities with an Emphasis on Sea-Based Development

Step 1: Screening Sea-Based Development Entrepreneurial Opportunities
Through a review of theoretical foundations, 48 entrepreneurial opportunities were identified across six dimensions. Following the evaluation, the designed questionnaire was validated. This questionnaire was then distributed to eight academic experts. After collecting the responses, the top 20 highest scoring factors were selected as the basis for analysis.

Step 2: Content Validity of the Indicators
The content validity was calculated using two criteria: CVR (content validity ratio) and CVI (content validity index). In this study, a panel of eight experts was utilized to evaluate and confirm the validity of the measures.

Content Validity Ratio (CVR) Index
The content validity ratio (CVR), developed by Lawshe, was used to assess the validity of the indicators. To calculate this index, expert judgments were collected from specialists in the relevant content domain. The experts were asked to evaluate each indicator based on a three-point Likert scale:

1. “The item is essential.”
2. “The item is useful but not essential.”
3. “The item is not necessary.”

Minimum Acceptable CVR Value Based on Expert Panel Size
The minimum acceptable content validity ratio (CVR) value depends on the number of experts evaluating the indicators, as determined by Lawshe’s reference table:

- Any indicator with a calculated CVR below the threshold (corresponding to the number of evaluating experts) must be excluded from the test.
- Such exclusion is necessary because these indicators fail to meet the required content validity standard (Table 2.44).

Table 2.45 presents the calculated content validity ratio (CVR) values for all measured criteria.

Content Validity Index (CVI)
To assess content validity, the Waltz and Bausell method is employed. In this approach, experts evaluate the “relevance” of each item using a four-point Likert

Table 2.44 Content validity ratio (CVR) index minimum acceptable CVR values based on number of expert raters

Number of experts	CVR value	Number of experts	CVR value	Number of experts	CVR value
5	0.99	11	0.59	25	0.37
6	0.99	12	0.56	30	0.33
7	0.99	13	0.54	35	0.31
8	0.75	14	0.51	40	0.29
9	0.78	15	0.49		
10	0.62	20	0.42		

scale. The experts rate the relevance of each item according to their judgment as follows:

1 = "Not relevant".
2 = "Requires major revision".
3 = "Relevant but needs minor revision".
4 = "Completely relevant"

$$CVI = \frac{Number\ of\ specialists\ evaluating\ the\ item\ as\ relevant\ (score \geq 3)}{Total\ number\ of\ content\ \exp erts\ participating\ in\ the\ validation}$$

The minimum acceptable value for the content validity index (CVI) is 0.75. Any item with a CVI score below 0.79 should be eliminated. Consequently, the identified factors along with their corresponding components are as follows (Table 2.46):

2.8.1.2 Prioritizing Main Criteria Using the Best-Worst Method (BWM)

Step 1: Identifying the Best and Worst Criteria
In this initial step, the decision-maker selects the best (most important) and worst (least important) criteria *without* conducting any pairwise comparisons.

Best and worst criteria:

$w_b = c_2$

$w_w = c_4$

In the **step 2**, the priority of the best criterion relative to the other criteria is determined using numbers between 1 and 9. The criterion "ecotourism and tourism jobs" has been selected as the best criterion, and its priority over the other criteria is determined. The priority of the ecotourism and tourism jobs criterion over the other criteria, based on the geometric mean of the opinions of ten experts, is presented in Table 2.47.

Step 3: Prioritizing all Criteria Relative to the Worst Criterion
The priority of all criteria relative to the worst criterion is determined using a scale of 1 to 9. Based on Table 2.48, the criterion "marine sector research and academic jobs" has been identified as the worst criterion. The priorities of the other criteria relative to

Table 2.45 Calculated content validity ratio (CVR) values for research criteria

Criteria	Essential responses	CVR value	Validation status
Ship electrical engineering (chief electrician)	8	0.75	Accepted
Deck officer (ship captain)	8	0.75	Accepted
Engine room sailor (ship fitter)	5	0.55	Rejected
Deck sailor (boatswain)	5	0.55	Rejected
Galley sailor (ship's head chef)	8	0.75	Accepted
Navigation officer	8	0.75	Accepted
Marine communications and electronics engineering	5	0.55	Rejected
Marine engineering	5	0.55	Rejected
Professional diving	5	0.55	Rejected
Marine biology	5	0.55	Rejected
Maritime legal specialist	8	0.75	Accepted
Marine research scientist	8	0.75	Accepted
Aquatic veterinarian	5	0.55	Rejected
Marine environmental protection specialist	5	0.55	Rejected
Marine research institute establishment	5	0.55	Rejected
Maritime museum design	5	0.55	Rejected
Submarine exploration	5	0.55	Rejected
Marine resources management	5	0.55	Rejected
Seawater desalination for drinking, industrial and agricultural use	5	0.55	Rejected
Coastal and marine ecotourism	8	0.75	Accepted
Eco-park and geo-park	5	0.55	Rejected
Coastal hospitality and recreational/resort complexes	8	0.75	Accepted
Indigenous handicrafts	8	0.75	Accepted
Seafood culinary services for tourists	8	0.75	Accepted
Beach sports	8	0.75	Accepted
Water sports	8	0.75	Accepted
Agritourism	8	0.75	Accepted
Health and wellness tourism	8	0.75	Accepted
Recreational boat manufacturing	5	0.55	Rejected
Construction and equipping of recreational marinas	5	0.55	Rejected
Cargo shipping and export activities	5	0.55	Rejected
Passenger transport and recreational boating	8	0.75	Accepted
Wind and solar energy production	5	0.55	Rejected
Marine pharmaceuticals and medical industries	5	0.55	Rejected
Maritime trade management	5	0.55	Rejected
Maritime-based trade (multipurpose economic ports)	5	0.55	Rejected
Offshore industries	5	0.55	Rejected
Marine international trade consultants	8	0.75	Accepted

(continued)

Table 2.45 (continued)

Criteria	Essential responses	CVR value	Validation status
Forestry and rangeland management	8	0.75	Accepted
Livestock and poultry farming	8	0.75	Accepted
Agriculture and horticulture	8	0.75	Accepted
Fishing and fisheries	8	0.75	Accepted
Marine products processing	8	0.75	Accepted
Cage fish farming	5	0.55	Rejected

Table 2.46 Final research criteria and sub-criteria

Main criterion	Code	Sub-criteria
Maritime occupations	C1	Marine electrical engineering Deck officers Cook/steward Navigation Legal specialist Marine researcher
Ecotourism and tourism occupations	C2	Coastal and marine ecotourism Coastal hospitality and resort management Indigenous handicrafts Seafood culinary services for tourists Beach sports Water sports Agritourism Health and wellness tourism Passenger transport and recreational boating
Agricultural sector	C3	Forestry and rangeland management Livestock and poultry farming Agriculture and horticulture Fishing and fisheries Marine products processing
Marine research and studies occupations	C4	International maritime trade consultants Legal specialists Marine researchers

Table 2.47 Priority of the ecotourism and tourism jobs criterion over other criteria

Ecotourism and tourism jobs	Maritime industry jobs	Agricultural sector	Research and academic jobs	Marine sector	Ecotourism and tourism jobs
4.914	4.666	4.888	1.000	–	1.000

Table 2.48 Priority of criteria relative to the worst criterion

Main criterion	Priority relative to marine sector research and academic jobs
Marine sector research and academic jobs (worst criterion)	1
Ecotourism and tourism jobs	6.933
Maritime industry jobs	6.165
Agricultural sector	4.133

this worst criterion, calculated using the geometric mean of inputs from ten experts, are presented in Table 2.48.

By computing the geometric mean of the ten experts' evaluations and inputting the results into the LINGO software, the final prioritization of the criteria is derived.

Table 2.49 Weights of main criteria

Main criterion	Weight
Agriculture	0.522
Ecotourism and tourism-related occupations	0.590
Maritime occupations	0.555
Marine research and academic occupations	0.214

The optimal weight assigned to the criteria is the weight that fulfills the following relationship:

$$\frac{W_b}{W_j} = a_{bj}, \quad \frac{W_j}{W_w} = a_{jw}.$$

Under the non-negativity conditions and the weight-normalization constraint, the following optimization problem is derived:

$$\begin{aligned} &\min_j \max\left\{\left|\frac{W_b}{W_j} - a_{bj}, \quad \frac{W_j}{W_w} - a_{jw}\right|\right\} \\ &s.t. \\ &\sum W_j = 1 \\ &W_j \geq 0\, for\, all\, j \end{aligned}$$

The model can be formulated as:

$$\begin{aligned} &Min\ \zeta \\ &s.t. \\ &\left|\frac{W_b}{W_j} - a_{bj}\right| \leq \zeta, \\ &\left|\frac{W_j}{W_w} - a_{jw}\right| \leq \zeta, \\ &\sum W_j = 1, \\ &W_j \geq 0\, for\, all\, j. \end{aligned}$$

Considering the ratings provided by ten experts and applying the best-worst method (BWM) constraints, the criteria weights were obtained using LINGO software (Table 2.49).

In the best-worst method (BWM), the consistency ratio must satisfy $0 \leq CR \leq 1$. It is computed using the following formula:

$$CI = \frac{\zeta^* - n}{n - 1}.$$

Table 2.50 Final weight of research sub-criteria

Criterion	Symbol	Criterion weight	Sub-criterion	Sub-criterion weight	Final sub-criterion weight	Rank
Maritime jobs	C1	0.522	Ship electrical engineering	0.243	0.126	6
			Deck officer	0.0256	0.0133	21
			Kitchen sailor	0.0839	0.0437	16
			Navigation	0.0362	0.0695	17
			Legal expert	0.0395	0.0206	20
			Marine researcher	0.0495	0.0062	23
Ecotourism and tourism jobs	C2	0.590	Coastal and marine ecotourism	0.274	0.466	2
			Hotel management and coastal resorts (and related jobs)	0.356	0.605	1
			Local handicrafts	0.411	0.242	3
			Seafood cooking for tourists	0.162	0.276	4
			Beach sports	0.17	0.10	9
			Water sports	0.155	0.091	11
			Agricultural tourism	0.103	0.060	14
			Health tourism	0.169	0.099	10
			Passenger transport and recreational boating	0.144	0.084	12
Agricultural jobs	C3	0.555	Forestry and range-land management	0.101	0.183	8
			Livestock and poultry farming	0.202	0.112	7
			Agriculture and horticulture	0.263	0.145	5
			Fishing and fisheries	0.104	0.057	15
			Marine products processing	0.127	0.070	13
Marine research and studies jobs	C4	0.212	International maritime trade consultants and experts	0.0298	0.0063	22
			Legal expert	0.144	0.030	18
			Marine researcher	0.183	0.021	19

The computed consistency ratio (CR = 0.72) satisfies the $0 \leq CR \leq 1$ condition, confirming the acceptability of the experts' pairwise comparisons in the BWM framework.

The sub-criteria were prioritized by multiplying the main criteria weights (stage 1) with their respective sub-criteria weights. Results are shown in Table 2.50.

References

Abdali, H., Sahebi, H., & Pishvaee, M. (2021). The water-energy-food-land nexus at the sugarcane-to-bioenergy supply chain: A sustainable network design model. *Computers and Chemical Engineering, 145*, 107–122.

Amirteimoori, A. R., & Soufi, M. (2019). Provide a new targeting model in a centralized decision-making environment with a multi-component network structure. *Journal of Operational Research and Its Applications, 16*(1), 93–115.

Asgarpour, M. J. (2023). *Multiple criteria decision making*. University of Tehran Press.

Dinara, F., & Soufi, M. (2016). Designing a mathematical model for optimal insurance branch location using a combination of fuzzy multi-criteria decision-making methods. *Journal of New Research in Mathematics, 39*(8), 97–112.

Duckstein, L., & Sudarovszky, F. (1994). Distance base techniques in multi criterion decision making. In J. J. Bogardi & H. P. Nachtnebel (Eds.), *Multi criterion decision analysis in water resources management*. UNESCO.

Fuller, R., & Carlsson, C. (1996). Fuzzy multiple criteria decision making: Recent developments. *Fuzzy Sets and Systems, 78*, 139–153.

Hamies, Y. Y., & Hall, W. A. (1974). Multi-objectives in water resources systems analysis: The Surrogate worth trade off method. *Water Resources Research, 10*(4), 615–625.

Hwang, C., & Yoon, K. (1981). *Multiple attribute decision making—Methods and applications. A state of the art survey*. Springer-Verlag.

Jafarnejad, A., Soufi, M., & Bayati, A. (2014). Prioritizing Critical Barriers of Computerized Maintenance Management System (CMMS) by Fuzzy Multi Attribute Decision Making (F-MADM) (Using LFPP). *Kuwait Chapter of Arabian Journal of Business and Management Review, 4*(3), 11–27.

Keeney, R. L., & Raiffa, H. (1993). *Decision with multiple objectives*. Cambridge University Press.

Lee, H., & Tanaka, H. (1999). Fuzzy approximations with non-symmetric fuzzy parameters in fuzzy regression analysis. *Journal of the Operations Research Society of Japan, 42*, 98–112.

Mikhailov, M. (2003). *Models, methods, concepts and applications of the analytic hierarchy process* (pp. 121–123). Springer Science & Business Media.

Nishizawa, K., & Takahashi, I. (2008). Weighted and Logarithmic Least Square Methods for Mutual Evaluation Network System including AHP and ANP. *Journal of the Operations Research Society of Japan, 50*(2), 101–122.

Saaty, T. L. (1980). *The analytic hierarchy process*. McGraw Hill, New York.

Safari, H., & Soufi, M. (2014). Select a hypermarket location based on fuzzy multi criteria decision making (F-MCDM) techniques (hybrid of F-DELPHI, F-AHP, F-LLSM and F-PROMETHEE). *Journal of Business and Management Review, 4*(1), 76–95.

Sekitani, K. (2000). Frobenius' theorem and its related topics: estimating the principal eigenvalue of a positive matrix including uncertain data. *ISM Reports on Statistical Computing, 125*, 29–46.

Sekitani, K., & Yamaki, N. (1999). A logical interpretation for the eigenvalue method in AHP. *Journal of the Operations Research Society of Japan, 42*(2), 219–232.

Wee, J. C. N., & Alton, Y. K. C. (2013). The peculiarities of knowledge management processes in SMEs: the case of Singapore Available to Purchase. *Journal of Knowledge Management, 17*(6), 958–972.

Chapter 3
Mathematical Programming-Based Techniques in MCDM

Learning Objectives

Upon completion of this chapter, the reader will be able to:

1. **Understand** the fundamental principles of mathematical programming and its role as a foundation for advanced multi-criteria decision-making (MCDM) techniques.
2. **Differentiate** between key mathematical programming-based MCDM methods, including the least squares method (LSM), logarithmic least squares method (LLSM), and linear programming technique for multidimensional analysis of preference (LINMAP).
3. **Formulate** and solve optimization models for criteria weighting and alternative prioritization using linear priority programming (LPP) and its logarithmic variant (logarithmic LPP).
4. **Apply** the LINMAP technique to rank alternatives based on their distance from an ideal solution, incorporating decision-maker preferences through pairwise comparisons.
5. **Analyze** decision problems under uncertainty by formulating and interpreting models that integrate fuzzy logic with mathematical programming, such as logarithmic fuzzy priority programming (LFPP).
6. **Evaluate** the strengths, weaknesses, and appropriate application contexts of each technique by examining real-world case studies in areas like supply chain management and facility location.

A. Amirteimoori, M. Soufi, *Industrial Decision Analysis*, Multiple Criteria Decision Making, https://doi.org/10.1007/978-3-032-24299-0_3

3.1 Introduction

Mathematical programming-based techniques in multi-criteria decision-making (MCDM) serve as a structured and powerful approach to solving complex problems involving multiple conflicting or contradictory criteria. By integrating the principles of mathematical programming and multi-criteria decision-making methods, these techniques enable the analysis and prioritization of alternatives based on multiple objectives. Among the well-known methods in this field are the **least squares method (LSM), linear least squares method (LLSM), linear programming with fuzzy parameters (LPP), linear fractional programming with fuzzy parameters (LFPP)**, and **LINMAP**. These techniques are widely applied in various domains such as supply chain management, industrial engineering, economics, environmental sciences, and other research fields. This book explores the concepts, algorithms, and applications of these techniques, serving as a comprehensive reference for researchers and professionals in the field.

3.2 Least Squares Method (LSM)

The least squares method (LSM) was first introduced by Carl Friedrich Gauss in the early nineteenth century and independently by Adrien-Marie Legendre. Gauss used this method to predict the orbit of the asteroid Ceres. The method quickly gained traction in various fields such as astronomy, geodesy, and economics. Today, it is one of the most fundamental techniques in statistics and engineering.

3.2.1 Definition of the LSM Technique

The least squares method is an optimization technique used to find the best fit of a model to observed data. The goal is to minimize the sum of the squares of the differences between the predicted values by the model and the actual values.

3.2.2 Mathematical Formulation

For a linear model $y = ax + b$, the objective is to minimize the following error function:

$$S = \sum_{i=1}^{n} (y_i - (ax_i + b))^2$$

where:

y_i represents the actual values.

$ax_i + b$ represents the predicted values by the model.

n is the number of data points.

However, as explained in Chap. 2, in the AHP method, the decision-maker compares the criteria or alternatives in a pairwise manner. The results of these comparisons are recorded in a square matrix called the **pairwise comparison matrix**. If the matrix is fully consistent, the following relationship holds between the elements of the matrix:

$$a_{ij} = \frac{w_i}{w_j}$$

where:

a_{ij} is the element of the pairwise comparison matrix.

$w_i$and w_j are the weights of criteria i and j, respectively.

If the pairwise comparison matrix is fully consistent, the above relationship holds for all elements of the matrix, and this is referred to as **complete consistency**. This is the ideal case, but it rarely occurs in practice. However, in most cases, the pairwise comparison matrix is inconsistent, meaning the above relationship does not hold precisely. This inconsistency arises due to human errors in comparisons or inaccuracies in evaluations, which is referred to as **incomplete consistency**.

Since the pairwise comparison matrix is usually inconsistent, the goal of the AHP method is to find weights (w_i) that minimize the error relative to the pairwise comparison matrix. In other words, we aim to find weights that satisfy the relationship ($a_{ij} \approx w_i \div w_j$) as closely as possible. The error (***e***) is defined as the difference between the actual value $\boldsymbol{a_{ij}}$ and the predicted value $\boldsymbol{w_j w_i}$. The objective is to minimize the sum of these errors.

The connection between LSM and AHP lies in the fact that, in the least squares method (LSM), the goal is to minimize the sum of squared errors. For the pairwise comparison matrix, this error is defined as follows:

$$e = \sum_{i=0}^{n} \sum_{j=1}^{n} \left(a_{ij} - \frac{w_i}{w_j} \right)^2$$

where:

a_{ij} is the element of the pairwise comparison matrix.

$\frac{w_i}{w_j}$ is the predicted value based on the weights.

In summary, the least squares method (LSM) is based on the principle that the minimum error occurs when the sum of squared errors (the second moment of errors) is minimized. This method ensures that errors are evenly distributed across all comparisons. In the AHP method, the goal is to find weights that bring the pairwise comparison matrix as close as possible to a state of consistency. This is achieved by minimizing the errors. The least squares method serves as a mathematical tool for minimizing the sum of squared errors and helps us determine the optimal weights.

3.2.3 Application of the LSM in MCDM

The least squares method (LSM) is a powerful mathematical tool used in multi-criteria decision-making (MCDM) to solve optimization problems and find the best solutions by minimizing errors. Its applications in MCDM include, but are not limited to, the following:

a. **Optimization of weights in weighting methods:** In many MCDM methods, such as the analytic hierarchy process (AHP) or the analytic network process (ANP), the weights of criteria or alternatives need to be calculated. The least squares method (LSM) is used to minimize the discrepancy between actual values (e.g., the pairwise comparison matrix) and predicted values (based on the weights). As mentioned earlier, LSM is applied in AHP to find weights that bring the pairwise comparison matrix as close as possible to a state of consistency.
b. **Solving regression problems in MCDM:** In some MCDM problems, it is necessary to model the relationship between criteria and alternatives. The least squares method (LSM) is used to fit linear or nonlinear regression models to the data in order to find the best relationship between the variables. In linear regression analysis, LSM is employed to determine the coefficients that minimize the sum of squared errors.
c. **Data normalization:** In some MCDM methods, such as TOPSIS or VIKOR, the least squares method (LSM) can be used to improve the accuracy of weight calculations or to normalize data.
d. **Solving nonlinear problems in MCDM:** Some MCDM problems are inherently nonlinear and require more advanced methods for their solution. The least squares method (LSM) can be utilized as part of nonlinear optimization algorithms to find optimal solutions. In nonlinear programming problems, LSM is employed to minimize the nonlinear objective function.
e. **Sensitivity analysis and decision-making improvement:** The least squares method (LSM) can be used for sensitivity analysis of decision-making outcomes. By minimizing errors, the impact of changes in data or weights can be examined, leading to improved decision-making. In the sensitivity analysis of AHP, LSM is employed to investigate the effect of changes in the pairwise comparison matrix on the final weights.

3.2.4 Strengths and Weaknesses of the LSM Technique

Strengths:

a. **Simplicity**: The formulation and implementation are straightforward.
b. **Computational efficiency**: Calculations are fast and efficient.
c. **High accuracy**: Provides precise results for linear data.
d. **Interpretability**: The coefficients of the model are easily interpretable.

Weaknesses:

a. **Sensitivity to outliers**: Outliers can significantly affect the results.
b. **Linearity limitation**: It is not suitable for nonlinear data without transformations or alternative methods.
c. **Statistical assumptions**: Requires assumptions such as normality of errors.

3.3 LLSM (Logarithmic Least Squares Method)

The **logarithmic least squares method (LLSM)** is a mathematical optimization technique used to fit a logarithmic model to a set of data points. The goal of this method is to minimize the sum of the squared differences between the observed values and the values predicted by the logarithmic model. This approach is particularly useful when the relationship between variables is expected to follow a logarithmic trend. By transforming the data into a logarithmic scale, the method linearizes the relationship, making it easier to apply linear regression techniques. The LLSM is commonly used in fields such as economics, biology, and engineering, where data often exhibit exponential or multiplicative patterns that can be effectively modeled using logarithmic functions.

3.3.1 Applications of LLSM in MCDM

In the field of multi-criteria decision-making (MCDM), the logarithmic least squares method (LLSM) is used to solve problems where data exhibits nonlinear or logarithmic relationships. Some of the main applications of LLSM in MCDM include:

a. **Criteria weighting:** Determining the weights of criteria in decision-making problems.
b. **Alternative ranking:** Evaluating and ranking alternatives based on various criteria.
c. **Data consistency analysis:** Examining data consistency and reducing errors caused by inconsistencies.

The **LLSM (logarithmic least squares method)** is a powerful tool for solving **multi-criteria decision-making (MCDM)** problems, particularly when the data exhibits nonlinear or logarithmic relationships. By applying a logarithmic transformation to the data and utilizing the least squares method, this approach enhances the accuracy and efficiency of the analysis. However, it is essential to consider its limitations and ensure the use of high-quality data for reliable results.

In general, the use of logarithms in **LLSM (logarithmic least squares method)** is typically applicable in the following scenarios:

a. **When the relationship between variables is nonlinear and requires linearization:** Some nonlinear problems can be transformed into linear ones using logarithms. This simplifies calculations and facilitates the application of LLSM.
b. **When data has a wide range or contains outliers:** Logarithms can reduce the impact of outliers by moderating large variations in data values.
c. **When the relationship between variables is exponential or power-based:** In cases where the relationship between variables is exponential or follows a power law, logarithms can linearize the relationship.
d. **When there is a need to homogenize variance:** In some cases, the variance of data increases with increasing values (heteroscedasticity). Logarithms can stabilize the variance, making it more homogeneous and meeting the assumptions required for LLSM.

   By addressing these scenarios, the use of logarithms in LLSM enhances the method's effectiveness and broadens its applicability to various types of data and relationships.

3.3.2 Mathematical Formulation of LLSM

The objective of the **logarithmic least squares method (LLSM)** is to minimize the sum of the squared differences between the observed values (y_i) and the predicted values ($\widehat{y}_i$) by a logarithmic model. This model is defined as follows:

$$\widehat{y}_i = a \times \ln(x_i) + b$$

where:x_i and y_i are the observed input and output values, respectively.

a and b are the model parameters to be determined.

ln(x_i) is the natural logarithm (Napierian logarithm) of x_i.

The objective function of LLSM is defined as the sum of squared errors:

$$Min\ S = \sum_{i=0}^{n} (y_i - \widehat{y}_i)^2 = \sum_{i=0}^{n} (y_i - (a \ln(x_i) + b))^2$$

where:

n is the number of data points.

S is the sum of squared errors.

To find the parameters ***a*** and ***b***, partial derivatives with respect to ***a*** and ***b*** are taken, and the normal equations are derived as follows:

$$\sum_{i=0}^{n}(y_i - \widehat{y}_i) = \sum_{i=0}^{n}(y_i - (a\ln(x_i) + b)).\ln(x_i) = 0$$

$$\sum_{i=0}^{n}(y_i - \widehat{y}_i) = \sum_{i=0}^{n}(y_i - (a\ln(x_i) + b)) = 0.$$

3.3.3 *Steps to Implement the LLSM Method*

The steps to implement the **logarithmic least squares method (LLSM)** are as follows:

a. **Data collection:** Gather data related to the criteria and alternatives.
b. **Logarithmic transformation:** Transform the data into a logarithmic scale to linearize the relationship between variables.
c. **Model fitting:** Use the least squares method to fit the logarithmic model.
d. **Error calculation:** Compute the sum of squared differences between the observed and predicted values.
e. **Optimization:** Minimize the errors to find the best-fitting model.

3.3.4 *Strengths and Weaknesses of the LLSM Method*

Strengths:

a. Ability to model nonlinear and logarithmic relationships.
b. Simplicity and applicability to complex problems.
c. High efficiency in analyzing data with exponential or multiplicative patterns.

Weaknesses:

a. Requires high-quality and error-free data.
b. Sensitivity to outliers.
c. Limited applicability to problems where the relationship between variables is not linear.

3.3.5 *Practical Differences Between LSM and LLSM Techniques in MCDM*

1. Scope of Application

LSM:

a. Can be used for both linear and nonlinear problems.
b. In nonlinear problems, LSM is employed to minimize the sum of squared errors in nonlinear functions.

LLSM:

a. Applicable only to linear problems.
b. In problems where the relationship between variables is linear, LLSM is used to find the best linear fit.

2. Computational Complexity

LSM:

a. Due to its ability to solve nonlinear problems, it involves more complex computations.
b. May require advanced algorithms such as numerical methods or nonlinear optimization techniques.

LLSM:

a. Simpler and faster computations, as it deals only with linear problems.
b. Typically uses straightforward matrix methods or solving linear equations.

3. Flexibility

LSM:

a. More flexible and can model more complex relationships between criteria and alternatives.
b. Suitable for problems requiring nonlinear or polynomial modeling.

LLSM:

a. Less flexible and suitable only for modeling linear relationships.
b. Best suited for problems where the relationship between variables is linear.

4. Accuracy and Consistency.

LSM:

a. In nonlinear problems, it may require more rigorous convergence and accuracy checks due to computational complexity.
b. May have multiple local solutions, necessitating advanced algorithms to find the global solution.

LLSM:

a. In linear problems, it usually provides a unique and precise solution.
b. Due to simpler computations, the accuracy of results is generally high.

5. Application in MCDM.

LSM:

a. Used in methods such as AHP, ANP, or TOPSIS, where nonlinear modeling or improving matrix consistency may be required.
b. Suitable for sensitivity analysis and enhancing decision-making in more complex problems.
c. In AHP, LSM is used to minimize errors arising from inconsistencies in pairwise comparison matrices.

LLSM:

a. Used in methods where the relationship between criteria is linear (e.g., certain types of linear regression in MCDM).
b. Suitable for simpler, linear problems in MCDM.

This comparison highlights the key differences in the application, complexity, flexibility, and suitability of LSM and LLSM techniques in the context of multi-criteria decision-making (MCDM).

3.4 Linear Priority Programming (LPP)

The LPP technique, or linear priority programming, is a mathematical optimization method used in multi-attribute decision-making (MADM) to calculate the weights of criteria or the priorities of alternatives. This method employs linear programming (LP) to solve decision-making problems involving multiple criteria and alternatives.

LPP is a linear programming method aimed at determining the weights of criteria or the priorities of alternatives in such a way that an objective function (such as minimizing error or maximizing consistency) is optimized. This method is widely used in multi-attribute decision-making (MADM), particularly in situations where data is precise and deterministic.

3.4.1 Applications of LPP in MADM

The application of LPP in MADM is extensive as a mathematical optimization method. This method is specifically used to determine the weights of criteria and prioritize alternatives in multi-criteria decision-making problems. Below are some of the most important applications of LPP in MADM:

1. **Determining criterion weights:**
   LPP is used to calculate the weights of criteria in MADM problems. These weights represent the relative importance of each criterion in the decision-making

process. Using LPP, weights can be calculated in a way that maximizes consistency between criteria and alternatives or minimizes errors.

2. **Prioritizing alternatives:**
   LPP helps decision-makers prioritize various alternatives based on multiple criteria. By optimizing the objective function (such as minimizing error or maximizing consistency), this method selects the best alternative based on the calculated weights.
3. **Solving problems with deterministic data:**
   LPP is particularly useful in situations where data is precise and deterministic. This method is suitable for problems where input information is available in quantitative and precise form.
4. **Improving consistency in decision-making:**
   LPP enhances the consistency between criteria and alternatives by optimizing the objective function. This helps decision-makers make more logical and coherent decisions.
5. **Applications in real-world problems:**
   LPP is applied in various fields such as supply chain management, supplier selection, project evaluation, human resource management, and strategic planning. For example:

   a. Supplier selection: Determining the best supplier based on criteria such as cost, quality, and on-time delivery.
   b. Project evaluation: Prioritizing projects based on criteria such as profitability, risk, and execution time.
   c. Human resource management: Evaluating and selecting employees based on criteria such as skills, experience, and performance.

6. **Reducing errors in decision-making:**
   LPP increases the accuracy and reliability of results by minimizing errors in the decision-making process. This is particularly useful in complex and multi-criteria problems.
7. **Flexibility in modeling:**
   LPP allows for the modeling of complex problems with multiple criteria and alternatives. This method provides the flexibility needed to incorporate constraints and specific conditions of each problem.
8. **Use in combination with other MADM methods:**
   LPP can be used as a complementary method alongside other multi-criteria decision-making techniques such as AHP (analytic hierarchy process) or TOPSIS (technique for order of preference by similarity to ideal solution) to provide more accurate and reliable results.

Conclusion

LPP is a powerful and practical method in MADM that helps decision-makers make optimal decisions by determining criterion weights and prioritizing alternatives. This method is particularly efficient in problems with deterministic data and complex conditions.

3.4.2 Mathematical Formulation of LLSM

The mathematical formulation of the **LPP** method for solving **MADM** problems is generally expressed as follows. This method uses linear programming to determine the weights of criteria or the priorities of alternatives. The general formulation of LPP is as follows:

$$Min Z = \sum_{i=0}^{n} w_i d_i$$

Or

$$Max Z = \sum_{i=0}^{n} w_i c_i$$

where:

Z is the objective function, which can either minimize error (d_i) or maximize consistency (c_i).
w_i is the weight of the ith criterion.
d_i represents the deviation or error for the $i$ith criterion.
c_i represents the consistency or alignment for the $i$ith criterion.
Constraints:

$$\sum_{i=0}^{n} w_i = 1$$
$$w_i \geq 0 \forall i$$

where:
w_i is the weight of the ith criterion and must be nonnegative.
The sum of the weights must equal 1 to ensure normalization.
Additional constraints (if required):
$A \cdot w \leq b.$
$C \cdot w = d.$
where:
A and C are matrices of constraint coefficients.
b and d are vectors of the right-hand side of the constraints.

Explanation

- The **objective function** can be defined as either minimizing the sum of errors or maximizing the consistency between criteria and alternatives.
- The **constraints** include the normalization of weights and other conditions (such as linear constraints).
- The **weights** (w_i) represent the relative importance of each criterion in the decision-making process.

This is the general formulation of LPP, which may vary depending on the type of MADM problem and the input data.

3.5 Logarithmic Priority Programming (LPP)

LPP is an advanced optimization technique used in **multi-criteria decision-making (MCDM)**. It is a mathematical approach that combines **logarithmic functions** with **priority programming** to determine the weights of criteria and prioritize alternatives in decision-making problems. LPP is particularly useful when dealing with complex decision-making scenarios where multiple criteria need to be evaluated simultaneously, and the relationships between criteria and alternatives are nonlinear.

3.5.1 Key Features of Logarithmic Priority Programming (LPP)

a. **Logarithmic transformation:**
   LPP uses logarithmic functions to transform and normalize the values of criteria and alternatives. This helps in handling nonlinear relationships and scaling issues in the data.
b. **Priority programming:**
   The method focuses on determining the priority or weight of each criterion based on its importance in the decision-making process. It aims to optimize an objective function, such as minimizing inconsistency or maximizing alignment with decision-maker preferences.
c. **Mathematical optimization:**
   LPP formulates the decision problem as a mathematical optimization model, often using linear or nonlinear programming techniques to solve for the optimal weights and priorities.

3.5.2 Formulation of LPP

The general formulation of LPP involves the following steps:

a. **Define the objective function**:
   The objective function is typically designed to minimize inconsistency or maximize consistency in the decision-making process. For example:

$$Min\ Z = \sum_{i=0}^{n} \log(w_i)\, d_i$$

   where:
   w_i is the weight of the ith criterion.
   d_i represents the deviation or inconsistency for the ith criterion.

b. **Apply logarithmic transformation:**
Logarithmic functions are applied to the weights or criteria values to handle nonlinearity and improve scaling. For example:

$$\log(w_i) = \log(a_i)$$

where a_i represents the raw value of the ith criterion.

c. **Include constraints:**
Constraints are added to ensure that the weights are normalized and nonnegative. For example:

$$\sum_{i=0}^{n} w_i = 1 \\ w_i \geq 0 \forall i.$$

d. **Solve the optimization problem:**
The optimization problem is solved using linear or nonlinear programming techniques to determine the optimal weights (w_i) and priorities.

Applications of LPP in MCDM

a. **Weight determination:**
LPP is used to calculate the weights of criteria in MCDM problems, reflecting their relative importance.

b. **Alternative prioritization:**
The method helps prioritize alternatives based on their performance across multiple criteria.

c. **Handling nonlinear relationships:**
LPP is particularly useful when the relationships between criteria and alternatives are nonlinear, as logarithmic transformations can linearize these relationships.

d. **Complex decision-making:**
LPP is applied in complex decision-making scenarios, such as resource allocation, project selection, and supply chain management.

3.5.3 Advantages of LPP

a. **Improved accuracy:**
Logarithmic transformations improve the accuracy of weight calculations by addressing scaling and nonlinearity issues.

b. **Flexibility:**
LPP can handle a wide range of decision-making problems, including those with complex and nonlinear criteria.

c. **Normalization:**
   The method ensures that weights are normalized and consistent with decision-maker preferences.
d. **Efficient optimization:**
   LPP uses mathematical optimization techniques to efficiently solve for optimal weights and priorities.

3.5.4 Conclusion

Logarithmic priority programming (LPP) is a powerful MCDM technique that combines logarithmic transformations with priority programming to determine optimal weights and priorities in decision-making problems. It is particularly effective in handling nonlinear relationships and complex decision-making scenarios, making it a valuable tool for decision-makers in various fields.

3.6 LFPP (Logarithmic Fuzzy Priority Programming)

Logarithmic fuzzy priority programming (LFPP) is an advanced method in the field of MADM that combines **fuzzy priority programming** and **logarithmic functions**. This method is designed to solve decision-making problems under uncertainty and with fuzzy data. LFPP uses logarithmic functions to improve accuracy and flexibility in calculating weights and prioritizing alternatives.

3.6.1 Key Concepts of LFPP

a. **Fuzzy:**
   In this method, data and criteria are expressed in fuzzy (uncertain) terms. This allows decision-makers to model uncertainty and ambiguity in information.
b. **Logarithmic functions:**
   Logarithmic functions are used to transform fuzzy values into crisp values. These functions help reduce computational complexity and increase accuracy in determining weights.
c. **Priority programming:**
   This aspect of the method focuses on determining criterion weights and prioritizing alternatives. The main goal is to optimize the objective function (such as minimizing error or maximizing consistency).

3.6.2 Formulation of LFPP Technique

The **logarithmic fuzzy priority programming (LFPP)** technique is one of the multi-criteria decision-making (MCDM) methods used to solve optimization problems with fuzzy objectives and fuzzy priorities. This technique combines fuzzy concepts and linear programming, utilizing logarithmic functions to manage uncertainty and fuzzy priorities.

Steps for Formulating LFPP

a. **Definition of decision variables and objective functions**:

- Let $x = (x_1, x_2, \ldots, x_n)$ be the vector of decision variables.
- Define the objective functions $f_k(x)$ *for* $k = 1, 2, \ldots, K$.

b. **Fuzzification of objective functions and constraints:**

- The objective functions and constraints are defined in fuzzy terms. For example, a fuzzy objective function is represented as $\widetilde{f}_k(x)$.
- Each fuzzy objective function is described by a membership function $\mu_{\tilde{f}_k}(x)$, which indicates the degree of membership.

c. **Definition of logarithmic functions for managing uncertainty:**

- Logarithmic functions are used to transform fuzzy values into crisp values. This helps reduce computational complexity.
- The logarithmic function for the membership function is defined as:

$$\ln \mu_{\tilde{f}_k}(x).$$

- This transformation allows the model to better handle uncertainty.

d. **Definition of the overall objective function:**

- The overall objective function is defined as a combination of the logarithmic functions of the membership functions:

$$Max\,Z = \sum\nolimits_{k=1}^{K} w_k \times \ln \mu_{\tilde{f}_k}(x)$$

where w_k represents the weights associated with each objective function.

e. **Application of constraints:**

- Fuzzy and crisp constraints are added to the model. For example:

$$\widetilde{g}_j(x) \leq \widetilde{b}_j \quad for j = 1, 2, \ldots, m.$$

- These constraints can also be managed using logarithmic functions.

f. **Solving the optimization model:**

- The optimization model is solved using numerical methods or optimization software (such as Lingo, GAMS, or MATLAB).
- The optimal solution x^* and the values of the objective functions $f_k(x*)$ are computed.

3.6.3 Advantages and Disadvantages of LFPP

1. *Advantages*

   Advantages of LFPP:

   a. **Modeling uncertainty:**

   - The ability to model fuzzy data and uncertainty in decision-making.

   b. **High accuracy:**

   - The use of logarithmic functions increases accuracy in calculating weights and priorities.

   c. **Flexibility:**

   - The ability to adapt to complex and multi-criteria problems.

   d. **Reduced computational complexity:**

   - Logarithmic functions help simplify calculations and reduce problem-solving time.

2. *Disadvantages*

a. **Complex formulation:**

- The formulation of the LFPP model can be mathematically complex, especially when dealing with multiple fuzzy objectives and constraints. This complexity increases with the size and intricacy of the problem, making it challenging for decision-makers to implement without advanced expertise.

b. **Dependence on weight assignments:**

- The accuracy and reliability of the LFPP results heavily depend on the proper assignment of weights to the objective functions. Incorrect or subjective weight assignments can lead to biased or suboptimal solutions.

c. **Computational intensity:**

- Although logarithmic transformations simplify some aspects of the model, the overall computational process can still be intensive, particularly for large-scale problems with numerous variables and constraints.

d. **Difficulty in handling highly nonlinear problems:**

- LFPP is more suited for problems with moderate levels of nonlinearity. For highly nonlinear problems, the technique may struggle to provide accurate or efficient solutions.

e. **Sensitivity to membership function definitions:**

- The performance of LFPP is highly sensitive to the choice and definition of membership functions for fuzzy objectives and constraints. Poorly defined membership functions can lead to misleading results.

f. **Limited interpretability of results:**

- The use of logarithmic transformations and fuzzy logic can make the results less intuitive and harder to interpret for decision-makers who are not familiar with these concepts.

g. **Challenges in scaling:**

- LFPP may face scalability issues when applied to very large or complex decision-making problems, as the computational requirements and model complexity grow exponentially.

h. **Requirement for expert knowledge:**

- Implementing LFPP effectively requires a deep understanding of fuzzy logic, logarithmic transformations, and optimization techniques. This limits its accessibility to users without specialized knowledge.

i. **Potential for overfitting:**

- In some cases, the model may overfit the fuzzy data, leading to solutions that are overly tailored to the specific dataset and less generalizable to new or slightly different scenarios.

j. **Limited application scope:**

- While LFPP is effective for certain types of fuzzy multi-criteria decision-making problems, it may not be suitable for all problem types, particularly those requiring crisp or deterministic solutions.

This technique is applicable in MADM problems such as project selection, resource management, and strategic planning.

3.6.4 Applications of LFPP

a. **Decision-making under uncertainty:**

- LFPP is highly suitable for solving decision-making problems where data is imprecise or fuzzy.

b. **Determining criterion weights:**

- This method is used to calculate criterion weights in complex problems with fuzzy data.

c. **Prioritizing alternatives:**

- LFPP helps decision-makers prioritize alternatives based on multiple criteria and fuzzy data.

d. **Applications in various fields:**

- LFPP is applied in fields such as supply chain management, supplier selection, project evaluation, and strategic planning.

3.6.5 Conclusion

Logarithmic fuzzy priority programming (LFPP) is a powerful and flexible method for solving multi-attribute decision-making problems under uncertainty. By combining fuzzy concepts and logarithmic functions, this method enhances the accuracy and efficiency of the decision-making process.

3.7 LINMAP

(Linear Programming Technique for Multidimensional Analysis of Preference)

LINMAP is a multi-attribute decision-making (MADM) technique that employs linear programming to analyze preferences and rank alternatives. This method is grounded in pairwise comparisons and preference data, striving to determine the optimal alternative by minimizing the distance from an ideal solution (positive ideal) and maximizing the distance from a nonideal solution (negative ideal). Its robust mathematical foundation and ability to handle complex decision-making scenarios make it a valuable tool in the field of multi-criteria decision analysis.

3.7.1 Steps for Implementing LINMAP

1. ***Identifying criteria and alternatives:*** The relevant criteria and alternatives associated with the decision-making problem are determined.
2. ***Constructing the decision matrix*:** A decision matrix is formed, which includes the evaluation of alternatives based on the defined criteria.
3. ***Defining ideal and nonideal points:*** The ideal point (representing the best possible scenario) and the nonideal point (representing the worst possible scenario) are defined for each criterion.

4. ***Formulating the objective function:*** The objective function is formulated to minimize the distance of alternatives from the ideal point and maximize their distance from the nonideal point.
5. ***Solving the linear programming model:*** The LINMAP model is solved using linear programming techniques, and the alternatives are ranked accordingly. This structured approach ensures a systematic and mathematically rigorous process for multi-criteria decision-making, making LINMAP a reliable method for complex decision analysis.

3.7.2 Formulation of the LINMAP Technique in MADM

The LINMAP (linear programming technique for multidimensional analysis of preference) is one of the multi-attribute decision-making (MADM) methods used for ranking and selecting the best alternative among several options based on multiple criteria. This technique is based on linear programming (LP) and aims to identify the optimal alternative by utilizing the decision-maker's preference data.

Assume a set of alternatives $A = \{A_1, A_2, \ldots, A_m\}$ and a set of criteria $C = \{C_1, C_2, \ldots, C_n\}$. Each alternative A_i is evaluated with respect to criterion C_j, and its performance value is denoted as x_{ij}. The goal is to rank the alternatives based on the decision-maker's preferences.

Steps of Formulation

a. ***Constructing the decision matrix***:
   The decision matrix X is defined as follows:

$$X = \begin{bmatrix} x_{11} & x_{12} & \cdots & x_{14} \\ x_{21} & x_{22} & \cdots & x_{24} \\ \vdots & \vdots & \ddots & \vdots \\ x_{m1} & x_{m2} & \cdots & x_{mn} \end{bmatrix}$$

   where x_{ij} represents the performance value of alternative A_i with respect to criterion C_j.

b. ***Determining decision-maker's Preferences:*** *The decision-maker provides* their preferences in the form of pairwise comparisons (A_k, A_1), indicating that alternative A_k is preferred over alternative A_1.

c. ***Defining the objective function:*** The LINMAP technique aims to minimize the Euclidean distance between the alternatives and an ideal point. The objective function is defined as:

$$Min\ Z = \sum_{(k,l) \in P} (d_k - d_l)$$

where P is the set of pairwise preferences, d_k is the Euclidean distance of alternative A_k from the ideal point, and d_l is the Euclidean distance of alternative A_1 from the ideal point.

d. ***Constraints:***

The following constraints must be satisfied:

$$d_k \leq d_l \ \forall (k, l) \in P.$$

These constraints ensure that the preferred alternatives are closer to the ideal point.

e. ***Solving the linear programming problem:*** Using linear programming techniques, the optimization problem is solved, and the values of d_i for each alternative are calculated. The alternatives are then ranked based on their d_i values, with smaller d_i values indicating better rankings.

3.7.3 Applications of LINMAP

The LINMAP (linear programming technique for multidimensional analysis of preference) is a decision-making tool used in multi-attribute decision-making (MADM). Its primary application is to rank and select the best alternative from a set of options based on multiple criteria, using linear programming to model and solve the decision problem. Below are the key applications of LINMAP:

a. **Ranking and Selection of Alternatives.**

   - LINMAP is widely used to rank alternatives based on their performance across multiple criteria. It helps decision-makers identify the best option by minimizing the distance of alternatives from an ideal point.
   - **Example**: Selecting the best supplier, project, or investment option based on cost, quality, and delivery time.

b. **Preference-Based Decision-Making.**

   - LINMAP incorporates the decision-maker's preferences in the form of pairwise comparisons (e.g., "alternative A is preferred over alternative B"). This makes it highly useful in scenarios where subjective judgments play a significant role.
   - **Example**: Choosing a location for a new facility based on stakeholder preferences.

c. **Resource Allocation.**

   - LINMAP can be applied to allocate limited resources optimally by evaluating alternatives against multiple criteria and selecting the most efficient option.
   - **Example**: Allocating budgets to different departments or projects based on performance metrics.

d. **Performance Evaluation.**

- LINMAP is used to evaluate the performance of systems, organizations, or processes by comparing them against an ideal benchmark.
- **Example**: Assessing the performance of different branches of a company based on sales, customer satisfaction, and operational efficiency.

e. **Product Design and Development.**

- In engineering and product design, LINMAP helps evaluate design alternatives based on technical, economic, and customer satisfaction criteria.
- **Example**: Selecting the best design for a new product based on cost, functionality, and user preferences.

f. **Environmental and Sustainability Decisions.**

- LINMAP is applied in environmental decision-making to evaluate alternatives based on sustainability criteria such as environmental impact, cost, and social acceptability.
- **Example**: Choosing a renewable energy project based on environmental benefits, cost, and feasibility.

g. **Healthcare and Medical Decision-Making.**

- In healthcare, LINMAP can be used to evaluate treatment options, medical devices, or healthcare policies based on multiple criteria such as effectiveness, cost, and patient outcomes.
- **Example**: Selecting the best treatment plan for a patient based on efficacy, side effects, and cost.

h. **Supply Chain and Logistics.**

- LINMAP is applied in supply chain management to evaluate and select suppliers, transportation modes, or distribution strategies based on cost, reliability, and delivery time.
- **Example**: Choosing the best logistics provider based on cost, delivery speed, and service quality.

i. **Marketing and Consumer Choice.**

- LINMAP can be used to analyze consumer preferences and rank products or services based on attributes such as price, quality, and brand reputation.
- **Example**: Identifying the most preferred product features based on customer feedback.

j. **Strategic Planning.**

- LINMAP supports strategic decision-making by evaluating long-term plans or strategies against multiple objectives and constraints.
- **Example**: Selecting a business strategy based on profitability, risk, and market share.

3.7.4 Advantages of LINMAP

- **High accuracy:** By utilizing linear programming, it delivers more precise results.
- **Conceptual simplicity:** It is conceptually simple and easy to understand.
- **Incorporation of preferences**: LINMAP explicitly incorporates the decision-maker's preferences, making it highly adaptable to subjective decision-making scenarios.
- **Flexibility:** It can simultaneously handle both quantitative and qualitative criteria. Additionally, it can be applied to various problems with multiple criteria and alternatives.
- **Ideal point concept:** By minimizing the distance to the ideal point, it provides a clear and intuitive framework for decision-making.
- **Scalability**: It can be applied to problems involving a large number of alternatives and criteria.
- **Broad applicability:** It is widely used in various fields such as management, economics, and engineering.

3.7.5 Disadvantages of LINMAP

- **Dependence on crisp data:** LINMAP is designed for crisp and deterministic data, and it may not be suitable for fuzzy or imprecise data.
- **Dependence on assumptions:** LINMAP operates based on linear assumptions, which may not always hold true.
- **Dependence on pairwise comparisons:** The accuracy of LINMAP depends on the quality of the decision-maker's pairwise comparisons.
- **Computational complexity:** For very large problems, the linear programming formulation may become computationally intensive. Additionally, for large and complex problems, LINMAP calculations can be time-consuming.

  In summary, LINMAP is a versatile and powerful tool for solving complex decision-making problems across various fields, including engineering, business, healthcare, and environmental management. Its ability to incorporate preferences and handle multiple criteria makes it a valuable technique in MADM.

3.7.6 Similarities of LINMAP with Other MCDM Techniques

a. **Similarity with LPP (Logarithmic Priority Programming).**

   - Both methods utilize linear programming to solve multi-criteria decision-making problems.
   - Both methods focus on calculating weights and priorities.
   - Both methods are suitable for precise and deterministic data.

b. **Similarity with LSM (Least Squares Method).**

- Both methods employ optimization to minimize errors.
- Both methods aim to enhance accuracy and consistency in decision-making.

c. **Similarity with TOPSIS.**

- Both methods incorporate the concept of ideal and nonideal points.
- Both methods strive to minimize the distance of alternatives from the ideal point and maximize their distance from the nonideal point.

3.7.7 Differences Between LINMAP and Other Techniques

- **Difference with LPP:** LINMAP operates based on pairwise comparisons and ideal points, whereas LPP primarily focuses on weight calculation.
- **Difference with LSM: LINMAP employs linear programming, while LSM uses the minimization of the sum of squared errors.**
- **Difference with TOPSIS:** LINMAP utilizes linear programming to solve problems, whereas TOPSIS relies on simpler computations for ranking alternatives. The LINMAP technique is a powerful and widely used method in multi-attribute decision-making (MADM). The LINMAP technique is a robust and versatile approach in multi-attribute decision-making (MADM) that employs linear programming to analyze preferences and rank alternatives. Due to its precision and flexibility, this method finds extensive applications in various fields such as management, economics, and engineering. LINMAP shares many similarities with methods like LPP, LSM, and TOPSIS; however, it also exhibits key differences in its implementation and formulation.

3.7.8 Example for LINMAP Problem

Suppose a decision-maker has provided the following decision table for five alternatives and two criteria (Table 3.1):

Table 3.1 Decision table for five alternatives and two criteria

Alternatives	Criterion 1	Criterion 2
A_1	0	5
A_2	5	4
A_3	0	2
A_4	1	3
A_5	4	1

3.7.9 Steps to Solve the LINMAP Model

1. **Determining ordered pairs:**
   - The ordered pairs (K, L) are determined based on the decision-maker's preferences.
2. **Formulating constraints:**
   - A linear constraint is added to the model for each ordered pair.
3. **Adding the normalization constraint:**
   - A normalization constraint is added to ensure proper scaling of the weights.
4. **Solving the linear programming model:**
   - The model is solved using linear programming methods (such as the simplex method).
   - The weights of the criteria and the auxiliary variables are calculated.
5. **Ranking the alternatives:**
   - After calculating the weights, the alternatives are ranked based on their final scores.

3.7.10 Objective of the LINMAP Model

The primary objective of the LINMAP model is to minimize the sum of errors (or slack variables) so that the decision-maker's preferences are satisfied as accurately as possible. This is achieved by solving a linear programming problem in which:

- **Weights of the criteria** (w_1, w_2, ...) are calculated in such a way that the decision-maker's preferences are accurately modeled.
- **Slack variables** (Z_{KL}) represent the deviation from the preferences and should be as small as possible.

The LINMAP model consists of a series of linear equations and inequalities formulated based on the decision-maker's preferences and input data. The model is expressed as follows:

1. ***Constraints related to ordered pairs:***

- Each ordered pair (K, L) indicates that alternative K is preferred over alternative L.
- For each ordered pair, a linear constraint is defined to ensure that this preference is respected in the model.

Based on Table 3.2, the set of ordered pairs (Ω) is as follows:

Table 3.2 Composed matrix of paired comparisons of elements effective on implementing e-factory system

	Management	Human resource	Production strategies	Just in time production	Production technology	Total quality management	Organizational culture
Management	1	2.174	2.089	3.161	2.441	4.52	1.81
Human Resource	0.46	1	1.59	1.992	1.95	3.822	0.756
Production Strategies	0.478	0.625	1	2.314	0.812	3.37	0.52
Just in time Production	0.316	0.502	0.432	1	0.478	2.004	0.29
Production Technology	0.41	0.513	1.234	2.091	1	2.89	0.44
Total quality Management	0.221	0.261	0.296	0.498	0.344	1	0.366
Organizational culture	0.552	1.32	1.935	0.370	2.28	2.730	1

$\Omega = \{(1,2), (1,3), (1,4), (1,5), (2,3), (2,4), (2,5), (3,4), (3,5), (4,5)\}$.

- For example, the pair (1,2) indicates that alternative $\mathbf{A_1}$ is preferred over alternative $\mathbf{A_2}$ (where $\mathbf{A_K} = \mathbf{A_1}$ and $\mathbf{A_L} = \mathbf{A_2}$).

Using the explanations and relationships provided, the mathematical model for the example can be formulated as follows:

$25w_1 - 9w_2 - 10v_1 + 2v_2 + Z_{12} \geq 0$ (*First constraint for the first ordered pair*).

$21w_2 - 6v_2 + I_{31} \geq 0$ (*First constraint for the second ordered pair*).

⋮

⋮

⋮

$15w_1 - 8w_2 - 6v_1 + 4v_2 + Z_{45} \geq 0$ (*First constraint for the last ordered pair*).

In these constraints:

- w_1, w_2: *The weights of the criteria that need to be determined*
- v_1, v_2: *Auxiliary variables that can be positive, negative, or zero (free in sign)*
- Z_{KL}: *Slack variables used to satisfy the preference conditions.*

2. Normalization constraint:

- ***A normalization constraint is added to ensure that the weights are properly scaled.***
- ***In this example:***

$-20w_1 + 2w_2 + 8v_1 + 4v_2 = 1$ *(Normalization constraint).*

3. Sign conditions for variables:

- ***Weights and slack variables must be nonnegative:***

$$w_1, w_2, Z_{KL} \geq 0.$$

- ***Auxiliary variables*** v_1, v_2 ***can be positive, negative, or zero (free in sign):***

$v_1, v_2 =$ *free of sign.*

By solving the model, we obtain the following results:

1. $Z^* = 0.25$: The optimal value of the objective function Z is 0.25.
2. $Z^*_{12} = 0.25$: The optimal value of the slack variable Z_{12} is 0.25.
3. $Z^*_{KL} = 0$: The optimal value of the slack variable Z_{k1} is 0.
4. $W^* = (0.0277, 0.0554)$: The optimal weights for the criteria are $w_1 = 0.0277$ and $w_2 = 0.0554$.
5. $V^* = (0.0832, 0.1944)$: The optimal values for the auxiliary variables are $v_1 = 0.0832$ and $v_2 = 0.1944$.
6. $x^* = \left(\frac{v_1^*}{w_1^*}, \frac{v_2^*}{w_2^*}\right)$: The optimal solution $x*$ is calculated as the ratio of the optimal auxiliary variables v_1^* and v_2^* to the optimal weights w_1^* and w_2^*.
7. $x^* = (3, 30.5)$: The optimal solution values are $x_1^* = 3$ and $x_2^* = 30.5$.

8. $S^* = (13.5, 4.5, 13.5, 4.5, 13.5)$: The optimal slack variables S^* have the values 13.5, 4.5, 13.5, 4.5, and 13.5.

   Based on the values of S^*, the alternatives can be ranked as follows:

- Alternatives A_2 and A_4 are tied for the first rank.
- Alternatives A_1, A_3, and A_5 are tied for the second rank.

3.8 Case Study (1)

***Prioritize Barriers of e-Factories in Iran's Industries with Hybrid Multi-criteria Decision-Making (MCDM) Techniques (AHP-LLSM)* (Safari & Soufi, 2015)**

In this case study, the barriers to establishing e-factories in the industries of a country have been prioritized. For this purpose, a deterministic weight vector was utilized through the logarithmic least squares method (LLSM) and a pairwise comparison matrix within the framework of the analytic hierarchy process (AHP).

The Problem-Solving Process in This Study Is as Follows:

First Step

Based on studies related to computer-integrated manufacturing (CIM) and its dependent components, as well as identifying the roadmap for establishing e-factories in theoretical studies, the dimensions and indicators for creating e-factories were determined. Subsequently, the opinions of industry experts regarding the critical barriers and challenges in implementing e-factory systems were collected through questionnaires to prioritize them accordingly.

Second Step

To rank and prioritize these elements based on their interrelationships, the analytic hierarchy process (AHP) technique was employed. Accordingly, another questionnaire was designed, and its data was used as inputs for the logarithmic least squares method (LLSM). By prioritizing these barriers using the LLSM method, the challenges in establishing e-factories were ranked. The final weight of each indicator was calculated using the LINGO software. The group composite matrices of pairwise comparisons for each group at different levels of the AHP tree were calculated separately for each element. The overall composition of the elements is presented in Table 3.2.

Third Step

After ensuring the accuracy of the calculations, the total weight of the alternatives—which represent the barriers to implementing e-factory systems in industries—was calculated to prioritize them.

For the analytic hierarchy process (AHP) with complete information, the solution of the logarithmic least squares method (LLSM) is equivalent to that of the geometric mean method. This method inherently provides a positive solution, as the

Table 3.3 Prioritizing barriers effective on implementation of e-factory system with AHP-LLSM method

Criteria	Weight
Management	0.2870762
Organizational culture	0.1733387
Human resources	0.1689245
Production strategies	0.1218263
Production technology	0.1185071
Just in time production	0.832436E-01
Total quality management	0.470834E-01

exponential values obtained from the inverse transformation of logarithms are always positive (Table 3.3).

A portion of the LLSM coding in this study is provided below:

$$Min = \left((@log(w_1) - @log(w_2) - @log(2))^2\right) + \left((@log(w_1) - @log(w_3) - @log(0.3333))^2\right)$$
$$+ \left((@log(w_2) - @log(w_1) - @log(0.5))^2\right) + \left((@log(w_2) - @log(w3) - @log(4))^2\right)$$
$$+ \left((@log(w_3) - @log(w_1) - @log(3))^2\right) + \left((@log(w_3) - @log(w_2) - @log(0.25))^2\right);$$

.

.

.

$w_1 + w_2 + w_3 = 1;$

$w_1 \geq 0;$

$w_2 \geq 0;$

$w_3 \geq 0;$

.

.

.

3.9 Case Study (2)

***Select a Hypermarket Location Based on Fuzzy Multi-criteria Decision-Making (F-MCDM) techniques (Hybrid of F-Delphi, F-AHP, F-LLSM, and F-PROMETHEE)* (Safari & Soufi, 2014).**

This study was conducted for the selection of a hypermarket location. In this model, there are five criteria and six alternatives (locations). This paper addresses a facility location allocation problem using a factor rating system, as outlined below:

1. **Determining alternatives, criteria, and decision-makers:** Defining linguistic values and their corresponding fuzzy numbers and aggregating the decision-makers' assessments using F-Delphi and F-AHP.
2. **Calculating the average fuzzy weight and constructing the fuzzy decision matrix:** Utilizing F-LLSM and LINGO software.
3. **Constructing the fuzzy preference function:** Defining the multi-criteria preference index to determine the valued outranking relation and calculating the positive outranking flow ($\Phi^+(a)$) and the negative outranking flow ($\Phi^-(a)$) to pre-order the alternatives using F-PROMETHEE.
4. **Defuzzification:** Applying Chen's maxima and minima sets and ranking the alternatives using Excel.

   These steps comprehensively address the location selection problem using fuzzy and multi-criteria methods. Here, only the F-LLSM part is explained in detail.

   To calculate the weights of the criteria, the F-LLSM method and LINGO software were utilized. After collecting the data for each alternative and converting linguistic variables into quantitative values, Table 3.4 can be used to obtain the fuzzy pairwise comparison matrix (3.5). Subsequently, the weights of the criteria can be calculated using F-LLSM.

In this problem, there are six locations with five criteria that are to be prioritized in order to determine the optimal location for the establishment of a hypermarket (Table 3.5).

By implementing the coding in the LINGO software, a portion of which is provided in this section and the complete version of which is included in the appendix of this chapter, the output of Table 3.6 is obtained. These weights are utilized in the **PROMETHEE** technique:

Table 3.4 Linguistic variables for fuzzy group pairwise comparisons matrix

Linguistic variables	Triangular fuzzy numbers		
Very poor	0	0	2
Poor	1	2	3
Medium poor	2	3.5	5
Medium	4	5	6
Medium good	5	6.5	8
Good	7	8	9
Very good	8	10	10

Table 3.5 Fuzzy aggregated decision matrix

Aggregate decision matrix	C1			C2			C3			C4			C5		
A1	0.534	0.634	0.734	0.155	0.228	0.355	0.179	0.266	0.379	0.428	0.548	0.669	0.414	0.521	0.628
A2	0.228	0.371	0.514	0.300	0.395	0.559	0.048	0.088	0.266	0.169	0.278	0.386	0.348	0.448	0.548
A3	0.717	0.817	0.917	0.721	0.829	0.938	0.686	0.790	0.893	0.400	0.500	0.600	0.483	0.590	0.697
A4	0.569	0.671	0.786	0.348	0.448	0.548	0.393	0.516	0.638	0.438	0.553	0.669	0.303	0.369	0.524
A5	0.400	0.500	0.600	0.472	0.614	0.755	0.459	0.567	0.676	0.266	0.386	0.507	0.262	0.338	0.497
A6	0.466	0.593	0.721	0.286	0.428	0.569	0.576	0.679	0.783	0.252	0.360	0.469	0.200	0.255	0.400
Weight	0.090	0.172	0.303	0.293	0.417	0.541	0.469	0.583	0.697	0.466	0.591	0.717	0.631	0.724	0.831

Table 3.6 Weights of criteria for DM1 with F-LLSM by LINGO

	Criteria 1			Criteria 2			Criteria 3			Criteria 4			Criteria 5		
Weight	W_{11}	W_{12}	W_{13}	W_{21}	W_{22}	W_{23}	W_{31}	W_{32}	W_{33}	W_{41}	W_{42}	W_{43}	W_{51}	W_{52}	W_{53}
Value	0.22	0.27	0.31	0.25	0.31	0.36	0.13	0.17	0.22	0.68	0.78	0.96	0.08	0.09	0.10

$$
\begin{aligned}
min = & \left((@log(w_{11}) - @log(w_{23}) - @log(1))^2\right) + \left((@log(w_{12}) - @log(w_{22}) - @log(2))^2\right) \\
& + \left((@log(w_{13}) - @log(w_{21}) - @log(3))^2\right) + \left((@log(w_{11}) - @log(w_{23}) - @log(1))^2\right) \\
& + \left((@log(w_{12}) - @log(w_{22}) - @log(1))^2\right) + \left((@log(w_{13}) - @log(w_{21}) - @log(1))^2\right) \\
& + \left((@log(w_{11}) - @log(w_{23}) - @log(1))^2\right) + \left((@log(w_{12}) - @log(w_{22}) - @log(1))^2\right) \\
& + \left((@log(w_{13}) - @log(w_{21}) - @log(1))^2\right) + \left((@log(w_{11}) - @log(w_{23}) - @log(0.33))^2\right) \\
& + \left((@log(w_{12}) - @log(w_{22}) - @log(0.5))^2\right) + \left((@log(w_{13}) - @log(w_{21}) - @log(1))^2\right) \\
& + \left((@log(w_{11}) - @log(w_{23}) - @log(0.33))^2\right) + \left((@log(w_{12}) - @log(w_{22}) - @log(0.5))^2\right) \\
& + \left((@log(w_{13}) - @log(w_{21}) - @log(1))^2\right) + \left((@log(w) - @log(w_{33}) - @log(3))^2\right) \\
& + \left((@log(w_{12}) - @log(w_{32}) - @log(4))^2\right) + \left((@log(w_{13}) - @log(w_{31}) - @log(5))^2\right) \\
& + \left((@log(w_{11}) - @log(w_{33}) - @log(1))^2\right) + \left((@log(w_{12}) - @log(w_{32}) - @log(2))^2\right) \\
& + \left((@log(w_{13}) - @log(w_{31}) - @log(3))^2\right) + \left((@log(w_{11}) - @log(w_{33}) - @log(1))^2\right) \\
& + \left((@log(w_{12}) - @log(w_{32}) - @log(2))^2\right) + \left((@log(w_{13}) - @log(w_{31}) - @log(3))^2\right) \\
& + \left((@log(w_{11}) - @log(w_{33}) - @log(1))^2\right) + \left((@log(w_{12}) - @log(w_{32}) - @log(1))^2\right) \\
& + \left((@log(w_{13}) - @log(w_{31}) - @log(1))^2\right) + \left((@log(w_{11}) - @log(w_{33}) - @log(1))^2\right) \\
& + \left((@log(w_{12}) - @log(w_{32}) - @log(2))^2\right) + \left((@log(w_{13}) - @log(w_{31}) - @log(3))^2\right) \\
& + \ldots\ldots\ldots\ldots\ldots\ldots
\end{aligned}
$$

$$
\begin{aligned}
& w_{11} + w_{23} + w_{33} + w_{43} + w_{53} + w_{63} \geq 1; \\
& w_{21} + w_{13} + w_{33} + w_{43} + w_{53} + w_{63} \geq 1; \\
& w_{13} + w_{23} + w_{13} + w_{43} + w_{53} + w_{63} \geq 1; \\
& w_{41} + w_{23} + w_{33} + w_{13} + w_{53} + w_{63} \geq 1; \\
& w_{51} + w_{23} + w_{33} + w_{43} + w_{13} + w_{63} \geq 1; \\
& w_{61} + w_{23} + w_{33} + w_{43} + w_{53} + w_{13} \geq 1; \\
& . \\
& . \\
& . \\
& w_{11} + w_{13} + w_{21} + w_{23} + w_{31} + w_{33} + w_{41} + w_{43} + w_{51} + w_{53} + w_{61} + w_{63} = 2; \\
& w_{13} \geq w_{12}; \\
& w_{12} \geq w_{11}; \\
& w_{11} > 0; \\
& . \\
& . \\
& .
\end{aligned}
$$

3.10 Case Study (3)

Prioritizing Critical Barriers of Computerized Maintenance Management System (CMMS) by Fuzzy Multi-attribute Decision-Making (F-MADM) (Using LFPP) (*Jafarnejad et al.**, 2014*)

This study was conducted at Shoaeshargh Concrete Company, which specializes in producing a wide range of high-quality building components, including joists, concrete blocks, clinker blocks, foam blocks, ready-made concrete walls and fences, preshrunk concrete, mosaic, terrazzo, floor finishes, Badbor tiles, septic tanks, barbed wire columns, column heads, and ready-made concrete chairs and tables, among others. The company aims to enhance the quality of its products and achieve high standards by adopting advanced technologies.

The company operates 77 machines, including scales, lift trucks, mixer trucks, cranes, trucks, dumpers, concrete breakers, and similar equipment. To improve operational efficiency, the company seeks to computerize its maintenance management system (CMMS) to monitor and control its machinery effectively and reduce the costs associated with emergency repairs. As part of this initiative, the company aims to identify and prioritize the barriers to computerization based on expert opinions and take appropriate steps to establish the necessary infrastructure.

In this context, the study focuses on prioritizing the critical barriers to CMMS implementation using the **logarithmic fuzzy prioritization program (LFPP)**. The identified factors include:

- **F1:** Equipment availability.
- **F2:** Labor productivity.
- **F3:** Maintenance information.
- **F4:** Management support.
- **F5:** Inventory control.
- **F6:** Environmental controls.

The LFPP method involves determining the weights of each factor through pairwise comparison matrices. Each factor is compared relative to the others to establish its priority. The pairwise comparison matrix used for ranking these factors is presented in Table 3.7.

Based on the pairwise comparison matrix presented in Table 3.6, the model is formulated to represent the comparison matrix. This problem is solved using the **logarithmic fuzzy prioritization program (LFPP)**. To implement the LFPP method, the **LINGO** toolbox is utilized. The results obtained from solving the nonlinear programming problem using the LFPP algorithm are presented in Table 3.8.

Table 3.7 Pairwise comparison matrix

	F1	F2	F3	F4	F5	F6
F1	(1, 1, 1)	(6, 7, 8)	(3, 4, 5)	(0.50, 1.5, 2.5)	(0.20, 0.25, 0.33)	(6, 7, 8)
F2	(0.13, 0.14, 0.17)	(1, 1, 1)	(1, 2, 3)	(4, 5, 6)	(0.17, 0.20, 0.25)	(1, 2, 3)
F3	(0.11, 0.13, 0.14)	(0.20, 0.25, 0.33)	(1, 1, 1)	(5, 6, 7)	(4, 5, 6)	(0.33, 0.5, 1)
F4	(0.14, 0.17, 0.20)	(1, 1, 1)	(4, 5, 6)	(6, 7, 8)	(0.17, 0.20, 0.25)	(3, 4, 5)
F5	(0.17, 0.20, 0.25)	(1, 1, 1)	(3, 4, 5)	(0.17, 0.20, 0.25)	(1, 1, 1)	(0.13, 0.14, 0.17)
F6	(0.13, 0.14, 0.17)	(0.20, 0.25, 0.33)	(1, 1, 1)	(0.4, 0.67, 2)	(0.33, 0.5, 1)	(7, 8, 9)

Table 3.8 Weights of factors

Factor	F1	F2	F3	F4	F5	F6
Weight	0.207055	0.179459	0.201009	0.219562	0.098242	0.094673

According to Table 3.7, **management support (F4)** is identified as the most significant factor influencing creativity, with the highest weight of **0.219562**. The remaining factors are ranked in the following order of importance:

- **F4:** Management support.
- **F1:** Equipment availability.
- **F3:** Maintenance information.
- **F2:** Labor productivity.
- **F5:** Inventory control.
- **F6:** Environmental controls.

This problem has been solved using the following code in the LINGO software:

$$
\begin{aligned}
& min = \left((1-L)^2\right) + 1000^*\left(\left(y_{12}^2\right) + \left(z_{12}^2\right) + \left(y_{13}^2\right) + \left(z_{13}^2\right) + \left(y_{14}^2\right)\right. \\
& \quad \left. + \left(z_{14}^2\right) + \left(y_{23}^2\right) + \left(z_{23}^2\right) + \left(y_{24}^2\right) + \left(z_{24}^2\right) + \left(y_{34}^2\right) + \left(z_{34}^2\right)\right); \\
& \quad x_1 - x_2 - (L^*@log(2.5)) - @log(0.2) + y_{12} \geq 0; \\
& \quad -x_1 + x_2 - (L^*@log(2)) + @log(1) + z_{12} \geq 0; \\
& \quad x_1 - x_3 - (L^*@log(1/15)) - @log(1/6) + y_{13} \geq 0; \\
& \quad -x_1 + x_3 - (L^*@log(15/8)) + @log(0.75) + z_{13} \geq 0; \\
& \quad x_1 - x_4 - (L^*@log(7/3)) - @log(1/7) + y_{14} \geq 0; \\
& \quad -x_1 + x_4 - (L^*@log(9/5)) + @log(3/5) + z_{14} \geq 0; \\
& \quad x_2 - x_3 - (L^*@log(8/5)) - @log(0.5) + y_{23} \geq 0; \\
& \quad -x_2 + x_3 - (L^*@log(25/16)) + @log(5/4) + z_{23} \geq 0; \\
& \quad x_2 - x_4 - (L^*@log(14/9)) - @log(3/7) + y_{24} \geq 0; \\
& \quad -x_2 + x_4 - (L^*@log(1.5)) + @log(1) + z_{24} \geq 0; \\
& \quad x_3 - x_4 - (L^*@log(35/24)) - @log(4/7) + y_{34} \geq 0; \\
& \quad -x_3 + x_4 - (L^*@log(36/25)) + @log(6/5) + z_{34} \geq 0; \\
& \quad x_1 \geq 0; \\
& \quad x_2 \geq 0; \\
& \quad x_3 \geq 0; \\
& \quad x_4 \geq 0; \\
& \quad L \geq 0; \\
& \quad y_{12} \geq 0; \\
& \quad y_{13} \geq 0; \\
& \quad y_{14} \geq 0; \\
& \quad y_{23} \geq 0; \\
& \quad y_{24} \geq 0; \\
& \quad y_{34} \geq 0; \\
& \quad z_{12} \geq 0; \\
& \quad z_{13} \geq 0; \\
& \quad z_{14} \geq 0; \\
& \quad z_{23} \geq 0; \\
& \quad z_{24} \geq 0; \\
& \quad z_{34} \geq 0; \\
& \quad w_1 = (@exp(x_1))/(@exp(x_1) + @exp(x_2) + @exp(x_3) + @exp(x_4)); \\
& \quad w_2 = (@exp(x_2))/(@exp(x_1) + @exp(x_2) + @exp(x_3) + @exp(x_4)); \\
& \quad w_3 = (@exp(x_3))/(@exp(x_1) + @exp(x_2) + @exp(x_3) + @exp(x_4)); \\
& \quad w_4 = (@exp(x_4))/(@exp(x_1) + @exp(x_2) + @exp(x_3) + @exp(x_4));
\end{aligned}
$$

Appendix 1

(Coding to Solve the Problem Using the FLLSM Technique)

$$min = \left((@log(w_{11}) - @log(w_{23}) - @log(1))^2\right) + \left((@log(w_{12}) - @log(w_{22}) - @log(2))^2\right)$$
$$+ \left((@log(w_{13}) - @log(w_{21}) - @log(3))^2\right) + \left((@log(w_{11}) - @log(w_{23}) - @log(1))^2\right)$$
$$+ \left((@log(w_{12}) - @log(w_{22}) - @log(1))^2\right) + \left((@log(w_{13}) - @log(w_{21}) - @log(1))^2\right)$$
$$+ \left((@log(w_{11}) - @log(w_{23}) - @log(1))^2\right) + \left((@log(w_{12}) - @log(w_{22}) - @log(1))^2\right)$$
$$+ \left((@log(w_{13}) - @log(w_{21}) - @log(1))^2\right) + \left((@log(w_{11}) - @log(w_{23}) - @log(0.33))^2\right)$$
$$+ \left((@log(w_{12}) - @log(w_{22}) - @log(0.5))^2\right) + \left((@log(w_{13}) - @log(w_{21}) - @log(1))^2\right)$$
$$+ \left((@log(w_{11}) - @log(w_{23}) - @log(0.33))^2\right) + \left((@log(w_{12}) - @log(w_{22}) - @log(0.5))^2\right)$$
$$+ \left((@log(w_{13}) - @log(w_{21}) - @log(1))^2\right) + \left((@log(w_{11}) - @log(w_{33}) - @log(3))^2\right)$$
$$+ \left((@log(w_{12}) - @log(w_{32}) - @log(4))^2\right) + \left((@log(w_{13}) - @log(w_{31}) - @log(5))^2\right)$$
$$+ \left((@log(w_{11}) - @log(w_{33}) - @log(1))^2\right) + \left((@log(w_{12}) - @log(w_{32}) - @log(2))^2\right)$$
$$+ \left((@log(w_{13}) - @log(w_{31}) - @log(3))^2\right) + \left((@log(w_{11}) - @log(w_{33}) - @log(1))^2\right)$$
$$+ \left((@log(w_{12}) - @log(w_{32}) - @log(2))^2\right) + \left((@log(w_{13}) - @log(w_{31}) - @log(3))^2\right)$$
$$+ \left((@log(w_{11}) - @log(w_{33}) - @log(1))^2\right) + \left((@log(w_{12}) - @log(w_{32}) - @log(1))^2\right)$$
$$+ \left((@log(w_{13}) - @log(w_{31}) - @log(1))^2\right) + \left((@log(w_{11}) - @log(w_{33}) - @log(1))^2\right)$$
$$+ \left((@log(w_{12}) - @log(w_{32}) - @log(2))^2\right) + \left((@log(w_{13}) - @log(w_{31}) - @log(3))^2\right)$$
$$+ \left((@log(w_{11}) - @log(w_{43}) - @log(4))^2\right) + \left((@log(w_{12}) - @log(w_{42}) - @log(5))^2\right)$$
$$+ \left((@log(w_{13}) - @log(w_{41}) - @log(6))^2\right) + \left((@log(w_{11}) - @log(w_{43}) - @log(3))^2\right)$$
$$+ \left((@log(w_{12}) - @log(w_{42}) - @log(4))^2\right) + \left((@log(w_{13}) - @log(w_{41}) - @log(5))^2\right)$$
$$+ \left((@log(w_{11}) - @log(w_{43}) - @log(2))^2\right) + \left((@log(w_{12}) - @log(w_{42}) - @log(3))^2\right)$$
$$+ \left((@log(w_{13}) - @log(w_{41}) - @log(4))^2\right) + \left((@log(w_{11}) - @log(w_{43}) - @log(1))^2\right)$$
$$+ \left((@log(w_{12}) - @log(w_{42}) - @log(2))^2\right) + \left((@log(w_{13}) - @log(w_{41}) - @log(3))^2\right)$$
$$+ \left((@log(w_{11}) - @log(w_{43}) - @log(2))^2\right) + \left((@log(w_{12}) - @log(w_{42}) - @log(3))^2\right)$$
$$+ \left((@log(w_{13}) - @log(w_{41}) - @log(4))^2\right) + \left((@log(w_{11}) - @log(w_{53}) - @log(2))^2\right)$$
$$+ \left((@log(w_{12}) - @log(w_{52}) - @log(3))^2\right) + \left((@log(w_{13}) - @log(w_{51}) - @log(4))^2\right)$$
$$+ \left((@log(w_{11}) - @log(w_{53}) - @log(1))^2\right) + \left((@log(w_{12}) - @log(w_{52}) - @log(2))^2\right)$$
$$+ \left((@log(w_{13}) - @log(w_{51}) - @log(3))^2\right) + \left((@log(w_{11}) - @log(w_{53}) - @log(2))^2\right)$$
$$+ \left((@log(w_{12}) - @log(w_{52}) - @log(3))^2\right) + \left((@log(w_{13}) - @log(w_{51}) - @log(4))^2\right)$$
$$+ \left((@log(w_{11}) - @log(w_{53}) - @log(2))^2\right)$$

$$+\left((@log(w_{12})-@log(w_{52})-@log(3))^2\right)+\left((@log(w_{13})-@log(w_{51})-@log(4))^2\right)$$
$$+\left((@log(w_{11})-@log(w_{53})-@log(3))^2\right)+\left((@log(w_{12})-@log(w_{52})-@log(4))^2\right)$$
$$+\left((@log(w_{13})-@log(w_{51})-@log(5))^2\right)+$$

$$\left((@log(w_{11})-@log(w_{63})-@log(2))^2\right)+\left((@log(w_{12})-@log(w_{62})-@log(3))^2\right)$$
$$+\left((@log(w_{13})-@log(w_{61})-@log(4))^2\right)+\left((@log(w_{11})-@log(w_{63})-@log(2))^2\right)$$
$$+\left((@log(w_{12})-@log(w_{62})-@log(3))^2\right)+\left((@log(w_{13})-@log(w_{61})-@log(4))^2\right)$$
$$+\left((@log(w_{11})-@log(w_{63})-@log(2))^2\right)+\left((@log(w12)-@log(w_{62})-@log(3))^2\right)$$
$$+\left((@log(w_{13})-@log(w_{61})-@log(4))^2\right)+\left((@log(w_{11})-@log(w_{63})-@log(2))^2\right)$$
$$+\left((@log(w_{12})-@log(w_{62})-@log(3))^2\right)+\left((@log(w_{13})-@log(w_{61})-@log(4))^2\right)$$
$$+\left((@log(w_{11})-@log(w_{63})-@log(3))^2\right)+\left((@log(w_{12})-@log(w_{62})-@log(4))^2\right)$$
$$+\left((@log(w_{13})-@log(w_{61})-@log(5))^2\right)+\left((@log(w_{21})-@log(w_{13})-@log(0.33))^2\right)$$
$$+\left((@log(w_{22})-@log(w_{12})-@log(0.5))^2\right)+\left((@log(w_{23})-@log(w_{11})-@log(1))^2\right)$$
$$+\left((@log(w_{21})-@log(w_{13})-@log(1))^2\right)+\left((@log(w_{22})-@log(w_{12})-@log(1))^2\right)$$
$$+\left((@log(w_{23})-@log(w_{11})-@log(1))^2\right)+\left((@log(w_{21})-@log(w_{13})-@log(1))^2\right)$$
$$+\left((@log(w_{22})-@log(w_{12})-@log(1))^2\right)+\left((@log(w_{23})-@log(w_{11})-@log(1))^2\right)$$
$$+\left((@log(w_{21})-@log(w_{13})-@log(1))^2\right)+\left((@log(w_{22})-@log(w_{12})-@log(2))^2\right)$$
$$+\left((@log(w_{23})-@log(w_{11})-@log(3))^2\right)+\left((@log(w_{21})-@log(w_{13})-@log(1))^2\right)$$
$$+\left((@log(w_{22})-@log(w_{12})-@log(2))^2\right)+\left((@log(w_{23})-@log(w_{11})-@log(3))^2\right)+$$

$$\left((@log(w_{21})-@log(w_{33})-@log(1))^2\right)+\left((@log(w_{22})-@log(w_{32})-@log(2))^2\right)$$
$$+\left((@log(w_{23})-@log(w_{31})-@log(3))^2\right)+\left((@log(w_{21})-@log(w_{33})-@log(2))^2\right)$$
$$+\left((@log(w_{22})-@log(w_{32})-@log(3))^2\right)+\left((@log(w_{23})-@log(w_{31})-@log(4))^2\right)$$
$$+\left((@log(w_{21})-@log(w_{33})-@log(1))^2\right)+\left((@log(w_{22})-@log(w_{32})-@log(2))^2\right)$$
$$+\left((@log(w_{23})-@log(w_{31})-@log(3))^2\right)+\left((@log(w_{21})-@log(w_{33})-@log(1))^2\right)$$
$$+\left((@log(w_{22})-@log(w_{32})-@log(2))^2\right)+\left((@log(w_{23})-@log(w_{31})-@log(3))^2\right)$$
$$+\left((@log(w_{21})-@log(w_{33})-@log(2))^2\right)+\left((@log(w22)-@log(w_{32})-@log(3))^2\right)$$
$$+\left((@log(w_{23})-@log(w_{31})-@log(4))^2\right)+$$

$$
\begin{aligned}
&\left((@log(w_{21})-@log(w_{43})-@log(3))^2\right)+\left((@log(w_{22})-@log(w_{42})-@log(4))^2\right)\\
&+\left((@log(w_{23})-@log(w_{41})-@log(5))^2\right)+\left((@log(w_{21})-@log(w_{43})-@log(4))^2\right)\\
&+\left((@log(w_{22})-@log(w_{42})-@log(5))^2\right)+\left((@log(w_{23})-@log(w_{41})-@log(6))^2\right)\\
&+\left((@log(w_{21})-@log(w_{43})-@log(2))^2\right)+\left((@log(w_{22})-@log(w_{42})-@log(3))^2\right)\\
&+\left((@log(w_{23})-@log(w_{41})-@log(4))^2\right)+\left((@log(w_{21})-@log(w_{43})-@log(2))^2\right)\\
&+\left((@log(w_{22})-@log(w_{42})-@log(3))^2\right)+\left((@log(w_{23})-@log(w_{41})-@log(4))^2\right)\\
&+\left((@log(w_{21})-@log(w_{43})-@log(3))^2\right)+\left((@log(w_{22})-@log(w_{42})-@log(4))^2\right)\\
&+\left((@log(w_{23})-@log(w_{41})-@log(5))^2\right)+\left((@log(w_{21})-@log(w_{53})-@log(2))^2\right)\\
&+\left((@log(w_{22})-@log(w_{52})-@log(3))^2\right)+\left((@log(w_{23})-@log(w_{51})-@log(4))^2\right)\\
&+\left((@log(w_{21})-@log(w_{53})-@log(2))^2\right)+\left((@log(w_{22})-@log(w_{52})-@log(3))^2\right)\\
&+\left((@log(w_{23})-@log(w_{51})-@log(4))^2\right)+\left((@log(w_{21})-@log(w_{53})-@log(2))^2\right)\\
&+\left((@log(w_{22})-@log(w52)-@log(3))^2\right)+\left((@log(w_{23})-@log(w_{51})-@log(4))^2\right)\\
&+\left((@log(w_{21})-@log(w_{53})-@log(3))^2\right)+\left((@log(w_{22})-@log(w_{52})-@log(4))^2\right)\\
&+\left((@log(w_{23})-@log(w_{51})-@log(5))^2\right)+\left((@log(w_{21})-@log(w_{53})-@log(4))^2\right)\\
&+\left((@log(w_{22})-@log(w_{52})-@log(5))^2\right)+\left((@log(w_{23})-@log(w_{51})-@log(6))^2\right)\\
&+\left((@log(w_{21})-@log(w_{63})-@log(1))^2\right)+\left((@log(w_{22})-@log(w_{62})-@log(2))^2\right)\\
&+\left((@log(w_{23})-@log(w_{61})-@log(3))^2\right)+\left((@log(w_{21})-@log(w_{63})-@log(3))^2\right)\\
&+\left((@log(w_{22})-@log(w_{62})-@log(4))^2\right)+\left((@log(w_{23})-@log(w_{61})-@log(5))^2\right)\\
&+\left((@log(w_{21})-@log(w_{63})-@log(2))^2\right)+\left((@log(w_{22})-@log(w_{62})-@log(3))^2\right)\\
&+\left((@log(w_{23})-@log(w_{61})-@log(4))^2\right)+\left((@log(w_{21})-@log(w_{63})-@log(3))^2\right)\\
&+\left((@log(w22)-@log(w_{62})-@log(4))^2\right)+\left((@log(w_{23})-@log(w_{61})-@log(5))^2\right)\\
&+\left((@log(w_{21})-@log(w_{63})-@log(4))^2\right)+\left((@log(w_{22})-@log(w_{62})-@log(5))^2\right)\\
&+\left((@log(w_{23})-@log(w_{61})-@log(6))^2\right)+\left((@log(w_{31})-@log(w_{13})-@log(0.2))^2\right)\\
&+\left((@log(w_{32})-@log(w_{12})-@log(0.25))^2\right)+\left((@log(w_{33})-@log(w_{11})-@log(0.33))^2\right)\\
&+\left((@log(w_{31})-@log(w_{13})-@log(0.33))^2\right)+\left((@log(w_{32})-@log(w_{12})-@log(0.5))^2\right)\\
&+\left((@log(w_{33})-@log(w_{11})-@log(1))^2\right)+\left((@log(w_{31})-@log(w_{13})-@log(0.33))^2\right)\\
&+\left((@log(w_{32})-@log(w_{12})-@log(0.5))^2\right)+\left((@log(w_{33})-@log(w_{11})-@log(1))^2\right)\\
&+\left((@log(w_{31})-@log(w_{13})-@log(1))^2\right)+\left((@log(w_{32})-@log(w_{12})-@log(1))^2\right)\\
&+\left((@log(w_{33})-@log(w_{11})-@log(1))^2\right)+\left((@log(w_{31})-@log(w_{13})-@log(0.33))^2\right)\\
&+\left((@log(w_{32})-@log(w_{12})-@log(0.5))^2\right)+\left((@log(w_{33})-@log(w_{11})-@log(1))^2\right)+
\end{aligned}
$$

$$\left((@log(w_{31})-@log(w_{23})-@log(0.33))^2\right)+\left((@log(w_{32})-@log(w_{22})-@log(0.5))^2\right)$$
$$+\left((@log(w_{33})-@log(w_{21})-@log(1))^2\right)+\left((@log(w_{31})-@log(w_{23})-@log(0.25))^2\right)$$
$$+\left((@log(w_{32})-@log(w_{22})-@log(0.33))^2\right)+\left((@log(w_{33})-@log(w_{21})-@log(0.5))^2\right)$$
$$+\left((@log(w_{31})-@log(w_{23})-@log(0.33))^2\right)+\left((@log(w_{32})-@log(w_{22})-@log(0.5))^2\right)$$
$$+\left((@log(w_{33})-@log(w21)-@log(1))^2\right)+\left((@log(w_{31})-@log(w_{23})-@log(0.33))^2\right)$$
$$+\left((@log(w_{32})-@log(w22)-@log(0.5))^2\right)+\left((@log(w_{33})-@log(w_{21})-@log(1))^2\right)$$
$$+\left((@log(w_{31})-@log(w_{23})-@log(0.25))^2\right)+\left((@log(w_{32})-@log(w_{22})-@log(0.33))^2\right)$$
$$+\left((@log(w_{33})-@log(w21)-@log(0.5))^2\right)+\left((@log(w_{31})-@log(w_{43})-@log(3))^2\right)$$
$$+\left((@log(w_{32})-@log(w_{42})-@log(4))^2\right)+\left((@log(w_{33})-@log(w_{41})-@log(5))^2\right)$$
$$+\left((@log(w_{31})-@log(w_{43})-@log(2))^2\right)+\left((@log(w_{32})-@log(w_{42})-@log(3))^2\right)$$
$$+\left((@log(w_{33})-@log(w_{41})-@log(4))^2\right)+\left((@log(w31)-@log(w_{43})-@log(1))^2\right)$$
$$+\left((@log(w_{32})-@log(w_{42})-@log(2))^2\right)+\left((@log(w_{33})-@log(w_{41})-@log(3))^2\right)$$
$$+\left((@log(w_{31})-@log(w_{43})-@log(1))^2\right)+\left((@log(w_{32})-@log(w42)-@log(2))^2\right)$$
$$+\left((@log(w_{33})-@log(w_{41})-@log(3))^2\right)+\left((@log(w31)-@log(w_{43})-@log(1))^2\right)$$
$$+\left((@log(w_{32})-@log(w_{42})-@log(2))^2\right)+\left((@log(w_{33})-@log(w_{41})-@log(3))^2\right)$$
$$+\left((@log(w_{31})-@log(w_{53})-@log(1))^2\right)+\left((@log(w_{32})-@log(w_{52})-@log(2))^2\right)$$
$$+\left((@log(w_{33})-@log(w_{51})-@log(3))^2\right)+\left((@log(w_{31})-@log(w_{53})-@log(1))^2\right)$$
$$+\left((@log(w_{32})-@log(w_{52})-@log(1))^2\right)+\left((@log(w_{33})-@log(w_{51})-@log(1))^2\right)$$
$$+\left((@log(w_{31})-@log(w_{53})-@log(1))^2\right)+\left((@log(w_{32})-@log(w_{52})-@log(2))^2\right)$$
$$+\left((@log(w_{33})-@log(w51)-@log(3))^2\right)+\left((@log(w_{31})-@log(w_{53})-@log(2))^2\right)$$
$$+\left((@log(w_{32})-@log(w52)-@log(3))^2\right)+\left((@log(w_{33})-@log(w_{51})-@log(4))^2\right)$$
$$+\left((@log(w_{31})-@log(w_{53})-@log(2))^2\right)+\left((@log(w_{32})-@log(w_{52})-@log(3))^2\right)$$
$$+\left((@log(w_{33})-@log(w_{51})-@log(4))^2\right)+\left((@log(w_{31})-@log(w_{63})-@log(2))^2\right)$$
$$+\left((@log(w_{32})-@log(w_{62})-@log(3))^2\right)+\left((@log(w_{33})-@log(w_{61})-@log(4))^2\right)$$
$$+\left((@log(w_{31})-@log(w_{63})-@log(1))^2\right)+\left((@log(w_{32})-@log(w_{62})-@log(2))^2\right)$$
$$+\left((@log(w_{33})-@log(w_{61})-@log(3))^2\right)+\left((@log(w_{31})-@log(w_{63})-@log(1))^2\right)$$
$$+\left((@log(w_{32})-@log(w_{62})-@log(2))^2\right)+\left((@log(w_{33})-@log(w61)-@log(3))^2\right)$$
$$+\left((@log(w_{31})-@log(w_{63})-@log(2))^2\right)+\left((@log(w_{32})-@log(w_{62})-@log(3))^2\right)$$
$$+\left((@log(w_{33})-@log(w_{61})-@log(4))^2\right)+\left((@log(w_{31})-@log(w_{63})-@log(3))^2\right)$$
$$+\left((@log(w_{32})-@log(w_{62})-@log(4))^2\right)+\left((@log(w_{33})-@log(w61)-@log(5))^2\right)+$$

$$\left((@log(w_{41})-@log(w_{13})-@log(0.17))^2\right)+\left((@log(w_{42})-@log(w_{12})-@log(0.2))^2\right)$$
$$+\left((@log(w_{43})-@log(w_{11})-@log(0.25))^2\right)+\left((@log(w_{41})-@log(w_{13})-@log(0.2))^2\right)$$
$$+\left((@log(w_{42})-@log(w_{12})-@log(0.25))^2\right)+\left((@log(w_{43})-@log(w_{11})-@log(0.33))^2\right)$$
$$+\left((@log(w_{41})-@log(w_{13})-@log(0.25))^2\right)+\left((@log(w_{42})-@log(w_{12})-@log(0.33))^2\right)$$
$$+\left((@log(w_{43})-@log(w_{11})-@log(0.5))^2\right)+\left((@log(w_{41})-@log(w_{13})-@log(0.33))^2\right)$$
$$+\left((@log(w_{42})-@log(w_{12})-@log(0.5))^2\right)+\left((@log(w_{43})-@log(w_{11})-@log(1))^2\right)$$
$$+\left((@log(w_{41})-@log(w_{13})-@log(0.25))^2\right)+\left((@log(w_{42})-@log(w_{12})-@log(0.33))^2\right)$$
$$+\left((@log(w_{43})-@log(w_{11})-@log(0.5))^2\right)+\left((@log(w_{41})-@log(w_{23})-@log(0.2))^2\right)$$
$$+\left((@log(w_{42})-@log(w_{22})-@log(0.25))^2\right)+\left((@log(w_{43})-@log(w_{21})-@log(0.33))^2\right)$$
$$+\left((@log(w_{41})-@log(w_{23})-@log(0.16))^2\right)+\left((@log(w_{42})-@log(w_{22})-@log(0.2))^2\right)$$
$$+\left((@log(w_{43})-@log(w_{21})-@log(0.25))^2\right)+\left((@log(w_{41})-@log(w_{23})-@log(0.25))^2\right)$$
$$+\left((@log(w_{42})-@log(w_{22})-@log(0.33))^2\right)+\left((@log(w_{43})-@log(w_{21})-@log(0.5))^2\right)$$
$$+\left((@log(w_{41})-@log(w_{23})-@log(0.25))^2\right)+\left((@log(w_{42})-@log(w_{22})-@log(0.33))^2\right)$$
$$+\left((@log(w_{43})-@log(w_{21})-@log(0.5))^2\right)+\left((@log(w_{41})-@log(w_{23})-@log(0.2))^2\right)$$
$$+\left((@log(w_{42})-@log(w_{22})-@log(0.25))^2\right)+\left((@log(w_{43})-@log(w21)-@log(0.33))^2\right)+$$

$$\left((@log(w_{41})-@log(w_{33})-@log(0.2))^2\right)+\left((@log(w_{42})-@log(w_{32})-@log(0.25))^2\right)$$
$$+\left((@log(w_{43})-@log(w_{31})-@log(0.33))^2\right)+\left((@log(w_{41})-@log(w_{33})-@log(0.25))^2\right)$$
$$+\left((@log(w_{42})-@log(w_{32})-@log(0.33))^2\right)+\left((@log(w_{43})-@log(w_{31})-@log(0.5))^2\right)$$
$$+\left((@log(w_{41})-@log(w_{33})-@log(0.33))^2\right)+\left((@log(w_{42})-@log(w_{32})-@log(0.5))^2\right)$$
$$+\left((@log(w_{43})-@log(w_{31})-@log(1))^2\right)+\left((@log(w_{41})-@log(w_{33})-@log(0.33))^2\right)$$
$$+\left((@log(w_{42})-@log(w_{32})-@log(0.5))^2\right)+\left((@log(w_{43})-@log(w_{31})-@log(1))^2\right)$$
$$+\left((@log(w_{41})-@log(w_{33})-@log(0.33))^2\right)+\left((@log(w_{42})-@log(w_{32})-@log(0.5))^2\right)$$
$$+\left((@log(w_{43})-@log(w_{31})-@log(1))^2\right)+$$

$$\left((@log(w_{41})-@log(w_{53})-@log(0.25))^2\right)+\left((@log(w42)-@log(w_{52})-@log(0.33))^2\right)$$
$$+\left((@log(w_{43})-@log(w_{51})-@log(0.5))^2\right)+\left((@log(w_{41})-@log(w_{53})-@log(0.2))^2\right)$$
$$+\left((@log(w_{42})-@log(w_{52})-@log(0.25))^2\right)+\left((@log(w_{43})-@log(w_{51})-@log(0.33))^2\right)$$
$$+\left((@log(w_{41})-@log(w_{53})-@log(1))^2\right)+\left((@log(w_{42})-@log(w_{52})-@log(1))^2\right)$$
$$+\left((@log(w_{43})-@log(w_{51})-@log(1))^2\right)+\left((@log(w_{41})-@log(w_{53})-@log(1))^2\right)$$
$$+\left((@log(w_{42})-@log(w_{52})-@log(2))^2\right)+\left((@log(w_{43})-@log(w_{51})-@log(3))^2\right)$$
$$+\left((@log(w_{41})-@log(w_{53})-@log(2))^2\right)+\left((@log(w_{42})-@log(w_{52})-@log(3))^2\right)$$
$$+\left((@log(w_{43})-@log(w_{51})-@log(4))^2\right)+$$

$$\left((@log(w_{41})-@log(w_{63})-@log(0.33))^2\right)+\left((@log(w_{42})-@log(w_{62})-@log(0.5))^2\right)$$
$$+\left((@log(w_{43})-@log(w_{61})-@log(1))^2\right)+\left((@log(w_{41})-@log(w_{63})-@log(0.25))^2\right)$$
$$+\left((@log(w_{42})-@log(w_{62})-@log(0.33))^2\right)+\left((@log(w_{43})-@log(w_{61})-@log(0.5))^2\right)$$
$$+\left((@log(w_{41})-@log(w_{63})-@log(1))^2\right)+\left((@log(w_{42})-@log(w_{62})-@log(1))^2\right)$$
$$+\left((@log(w_{43})-@log(w_{61})-@log(1))^2\right)+\left((@log(w_{41})-@log(w_{63})-@log(1))^2\right)$$
$$+\left((@log(w_{42})-@log(w_{62})-@log(2))^2\right)+\left((@log(w_{43})-@log(w_{61})-@log(3))^2\right)$$
$$+\left((@log(w_{41})-@log(w_{63})-@log(2))^2\right)+\left((@log(w_{42})-@log(w_{62})-@log(3))^2\right)$$
$$+\left((@log(w_{43})-@log(w_{61})-@log(4))^2\right)+$$

$$\left((@log(w_{51})-@log(w_{13})-@log(0.25))^2\right)+\left((@log(w_{52})-@log(w_{12})-@log(0.33))^2\right)$$
$$+\left((@log(w_{53})-@log(w_{11})-@log(0.5))^2\right)+\left((@log(w_{51})-@log(w_{13})-@log(0.33))^2\right)$$
$$+\left((@log(w_{52})-@log(w_{12})-@log(0.5))^2\right)+\left((@log(w_{53})-@log(w_{11})-@log(1))^2\right)$$
$$+\left((@log(w_{51})-@log(w_{13})-@log(0.25))^2\right)+\left((@log(w_{52})-@log(w12)-@log(0.33))^2\right)$$
$$+\left((@log(w_{53})-@log(w_{11})-@log(0.5))^2\right)+\left((@log(w_{51})-@log(w_{13})-@log(0.25))^2\right)$$
$$+\left((@log(w_{52})-@log(w_{12})-@log(0.33))^2\right)+\left((@log(w_{53})-@log(w_{11})-@log(0.1))^2\right)$$
$$+\left((@log(w_{51})-@log(w_{13})-@log(0.2))^2\right)+\left((@log(w_{52})-@log(w_{12})-@log(0.33))^2\right)$$
$$+\left((@log(w_{53})-@log(w_{11})-@log(0.5))^2\right)+$$

$$\left((@log(w_{51}) - @log(w_{23}) - @log(0.25))^2\right) + \left((@log(w_{52}) - @log(w_{22}) - @log(0.33))^2\right)$$
$$+\left((@log(w_{53}) - @log(w_{21}) - @log(0.5))^2\right) + \left((@log(w_{51}) - @log(w_{23}) - @log(0.25))^2\right)$$
$$+\left((@log(w_{52}) - @log(w_{22}) - @log(0.33))^2\right) + \left((@log(w_{53}) - @log(w_{21}) - @log(0.5))^2\right)$$
$$+\left((@log(w_{51}) - @log(w_{23}) - @log(0.25))^2\right) + \left((@log(w_{52}) - @log(w_{22}) - @log(0.33))^2\right)$$
$$+\left((@log(w_{53}) - @log(w_{21}) - @log(0.5))^2\right) + \left((@log(w_{51}) - @log(w_{23}) - @log(0.2))^2\right)$$
$$+\left((@log(w_{52}) - @log(w22) - @log(0.25))^2\right) + \left((@log(w_{53}) - @log(w_{21}) - @log(0.33))^2\right)$$
$$+\left((@log(w_{51}) - @log(w_{23}) - @log(0.16))^2\right) + \left((@log(w_{52}) - @log(w_{22}) - @log(0.2))^2\right)$$
$$+\left((@log(w_{53}) - @log(w_{21}) - @log(0.25))^2\right) +$$

$$\left((@log(w_{51}) - @log(w_{33}) - @log(0.33))^2\right) + \left((@log(w_{52}) - @log(w_{32}) - @log(0.5))^2\right)$$
$$+\left((@log(w_{53}) - @log(w_{31}) - @log(1))^2\right) + \left((@log(w_{51}) - @log(w_{33}) - @log(1))^2\right)$$
$$+\left((@log(w_{52}) - @log(w_{32}) - @log(1))^2\right) + \left((@log(w_{53}) - @log(w_{31}) - @log(1))^2\right)$$
$$+\left((@log(w_{51}) - @log(w_{33}) - @log(0.33))^2\right) + \left((@log(w_{52}) - @log(w_{32}) - @log(0.5))^2\right)$$
$$+\left((@log(w_{53}) - @log(w31) - @log(1))^2\right) + \left((@log(w_{51}) - @log(w_{33}) - @log(0.25))^2\right)$$
$$+\left((@log(w_{52}) - @log(w_{32}) - @log(0.33))^2\right) + \left((@log(w_{53}) - @log(w_{31}) - @log(0.5))^2\right)$$
$$+\left((@log(w_{51}) - @log(w_{33}) - @log(0.25))^2\right) + \left((@log(w_{52}) - @log(w_{32}) - @log(0.33))^2\right)$$
$$+\left((@log(w_{53}) - @log(w_{31}) - @log(0.5))^2\right) +$$

$$\left((@log(w_{51}) - @log(w_{43}) - @log(2))^2\right) + \left((@log(w_{52}) - @log(w_{42}) - @log(3))^2\right)$$
$$+\left((@log(w_{53}) - @log(w_{41}) - @log(4))^2\right) + \left((@log(w_{51}) - @log(w_{43}) - @log(3))^2\right)$$
$$+\left((@log(w_{52}) - @log(w_{41}) - @log(4))^2\right) + \left((@log(w_{53}) - @log(w_{41}) - @log(5))^2\right)$$
$$+\left((@log(w_{51}) - @log(w_{43}) - @log(1))^2\right) + \left((@log(w_{52}) - @log(w_{42}) - @log(1))^2\right)$$
$$+\left((@log(w_{53}) - @log(w_{41}) - @log(1))^2\right) + \left((@log(w_{51}) - @log(w_{43}) - @log(0.33))^2\right)$$
$$+\left((@log(w_{52}) - @log(w_{42}) - @log(0.5))^2\right) + \left((@log(w_{53}) - @log(w_{41}) - @log(1))^2\right)$$
$$+\left((@log(w_{51}) - @log(w_{43}) - @log(0.25))^2\right) + \left((@log(w_{52}) - @log(w_{42}) - @log(0.33))^2\right)$$
$$+\left((@log(w_{53}) - @log(w_{41}) - @log(0.5))^2\right) +$$

$$\left((@log(w_{51})-@log(w_{63})-@log(1))^2\right)+\left((@log(w_{52})-@log(w_{62})-@log(2))^2\right)$$
$$+\left((@log(w_{53})-@log(w_{61})-@log(3))^2\right)+\left((@log(w_{51})-@log(w_{63})-@log(1))^2\right)$$
$$+\left((@log(w_{52})-@log(w_{62})-@log(2))^2\right)+\left((@log(w_{53})-@log(w_{61})-@log(3))^2\right)$$
$$+\left((@log(w_{51})-@log(w_{63})-@log(1))^2\right)+\left((@log(w_{52})-@log(w_{62})-@log(1))^2\right)$$
$$+\left((@log(w_{53})-@log(w_{61})-@log(1))^2\right)+\left((@log(w_{51})-@log(w_{63})-@log(1))^2\right)$$
$$+\left((@log(w_{52})-@log(w_{62})-@log(1))^2\right)+\left((@log(w_{53})-@log(w_{61})-@log(1))^2\right)$$
$$+\left((@log(w_{51})-@log(w_{63})-@log(1))^2\right)+\left((@log(w_{52})-@log(w_{62})-@log(1))^2\right)$$
$$+\left((@log(w_{53})-@log(w_{61})-@log(1))^2\right)+$$

$$\left((@log(w_{61})-@log(w_{13})-@log(0.25))^2\right)+\left((@log(w_{62})-@log(w_{12})-@log(0.33))^2\right)$$
$$+\left((@log(w_{63})-@log(w_{11})-@log(0.5))^2\right)+\left((@log(w_{61})-@log(w_{13})-@log(0.25))^2\right)$$
$$+\left((@log(w_{62})-@log(w_{12})-@log(0.33))^2\right)+\left((@log(w_{63})-@log(w_{11})-@log(0.5))^2\right)$$
$$+\left((@log(w_{61})-@log(w_{13})-@log(0.25))^2\right)+\left((@log(w_{62})-@log(w_{12})-@log(0.33))^2\right)$$
$$+\left((@log(w_{63})-@log(w_{11})-@log(0.5))^2\right)+\left((@log(w_{61})-@log(w_{13})-@log(0.25))^2\right)$$
$$+\left((@log(w_{62})-@log(w_{12})-@log(0.33))^2\right)+\left((@log(w_{63})-@log(w_{11})-@log(0.5))^2\right)$$
$$+\left((@log(w_{61})-@log(w_{13})-@log(0.2))^2\right)+\left((@log(w_{62})-@log(w_{12})-@log(0.25))^2\right)$$
$$+\left((@log(w_{63})-@log(w_{11})-@log(0.33))^2\right)+$$

$$\left((@log(w_{61})-@log(w_{23})-@log(0.33))^2\right)+\left((@log(w_{62})-@log(w_{22})-@log(0.5))^2\right)$$
$$+\left((@log(w_{63})-@log(w_{21})-@log(1))^2\right)+\left((@log(w_{61})-@log(w_{23})-@log(0.2))^2\right)$$
$$+\left((@log(w_{62})-@log(w_{22})-@log(0.25))^2\right)+\left((@log(w_{63})-@log(w_{21})-@log(0.33))^2\right)$$
$$+\left((@log(w_{61})-@log(w_{23})-@log(0.25))^2\right)+\left((@log(w_{62})-@log(w_{22})-@log(0.33))^2\right)$$
$$+\left((@log(w_{63})-@log(w_{21})-@log(0.5))^2\right)+\left((@log(w_{61})-@log(w_{23})-@log(0.2))^2\right)$$
$$+\left((@log(w_{62})-@log(w_{22})-@log(0.25))^2\right)+\left((@log(w_{63})-@log(w_{21})-@log(0.33))^2\right)$$
$$+\left((@log(w_{61})-@log(w_{23})-@log(0.16))^2\right)+\left((@log(w_{62})-@log(w_{22})-@log(0.2))^2\right)$$
$$+\left((@log(w_{63})-@log(w_{21})-@log(0.25))^2\right)+$$

$$\left((@log(w_{61})-@log(w_{33})-@log(0.25))^2\right)+\left((@log(w_{62})-@log(w_{32})-@log(0.33))^2\right)$$
$$+\left((@log(w_{63})-@log(w_{31})-@log(0.5))^2\right)+\left((@log(w_{61})-@log(w_{33})-@log(0.33))^2\right)$$
$$+\left((@log(w_{62})-@log(w_{32})-@log(0.5))^2\right)+\left((@log(w_{63})-@log(w_{31})-@log(1))^2\right)$$
$$+\left((@log(w_{61})-@log(w_{33})-@log(0.33))^2\right)+\left((@log(w_{62})-@log(w_{32})-@log(0.5))^2\right)$$
$$+\left((@log(w_{63})-@log(w_{31})-@log(1))^2\right)+\left((@log(w_{61})-@log(w_{33})-@log(0.25))^2\right)$$
$$+\left((@log(w_{62})-@log(w_{32})-@log(0.33))^2\right)+\left((@log(w_{63})-@log(w_{31})-@log(0.5))^2\right)$$
$$+\left((@log(w_{61})-@log(w_{33})-@log(0.25))^2\right)+\left((@log(w_{62})-@log(w_{32})-@log(0.33))^2\right)$$
$$+\left((@log(w_{63})-@log(w_{31})-@log(0.5))^2\right)+$$

$$\left((@log(w_{61})-@log(w_{43})-@log(1))^2\right)+\left((@log(w_{62})-@log(w_{42})-@log(2))^2\right)$$
$$+\left((@log(w_{63})-@log(w_{41})-@log(3))^2\right)+\left((@log(w_{61})-@log(w_{43})-@log(2))^2\right)$$
$$+\left((@log(w_{62})-@log(w_{42})-@log(3))^2\right)+\left((@log(w_{63})-@log(w_{41})-@log(4))^2\right)$$
$$+\left((@log(w_{61})-@log(w_{43})-@log(1))^2\right)+\left((@log(w_{62})-@log(w_{42})-@log(1))^2\right)$$
$$+\left((@log(w_{63})-@log(w_{41})-@log(1))^2\right)+\left((@log(w_{61})-@log(w_{43})-@log(0.33))^2\right)$$
$$+\left((@log(w_{62})-@log(w_{42})-@log(0.5))^2\right)+\left((@log(w_{63})-@log(w_{41})-@log(1))^2\right)$$
$$+\left((@log(w_{61})-@log(w_{43})-@log(0.25))^2\right)+\left((@log(w_{62})-@log(w_{42})-@log(0.33))^2\right)$$
$$+\left((@log(w_{63})-@log(w_{41})-@log(0.5))^2\right)+$$

$$\left((@log(w_{61})-@log(w_{53})-@log(0.33))^2\right)+\left((@log(w_{62})-@log(w_{52})-@log(0.5))^2\right)$$
$$+\left((@log(w_{63})-@log(w_{51})-@log(1))^2\right)+\left((@log(w_{61})-@log(w_{53})-@log(0.33))^2\right)$$
$$+\left((@log(w_{62})-@log(w_{52})-@log(0.5))^2\right)+\left((@log(w_{63})-@log(w_{51})-@log(1))^2\right)$$
$$+\left((@log(w_{61})-@log(w_{53})-@log(1))^2\right)+\left((@log(w_{62})-@log(w_{52})-@log(1))^2\right)$$
$$+\left((@log(w_{63})-@log(w_{51})-@log(1))^2\right)+\left((@log(w_{61})-@log(w_{53})-@log(1))^2\right)$$
$$+\left((@log(w_{62})-@log(w_{52})-@log(1))^2\right)+\left((@log(w_{63})-@log(w_{51})-@log(1))^2\right)$$
$$+\left((@log(w_{61})-@log(w_{53})-@log(1))^2\right)+\left((@log(w_{62})-@log(w_{52})-@log(1))^2\right)$$
$$+\left((@log(w_{63})-@log(w_{51})-@log(1))^2\right);$$

$$w_{11} + w_{23} + w_{33} + w_{43} + w_{53} + w_{63} \geq 1;$$

$$w_{21} + w_{13} + w_{33} + w_{43} + w_{53} + w_{63} \geq 1;$$

$$w_{31} + w_{23} + w_{13} + w_{43} + w_{53} + w_{63} \geq 1;$$

$$w_{41} + w_{23} + w_{33} + w_{13} + w_{53} + w_{63} \geq 1;$$

$$w_{51} + w_{23} + w_{33} + w_{43} + w_{13} + w_{63} \geq 1;$$

$$w_{61} + w_{23} + w_{33} + w_{43} + w_{53} + w_{13} \geq 1;$$

$$w_{13} + w_{21} + w_{31} + w_{41} + w_{51} + w_{61} \leq 1;$$

$$w_{23} + w_{11} + w_{31} + w_{41} + w_{51} + w_{61} \leq 1;$$

$$w_{33} + w_{21} + w11 + w_{41} + w_{51} + w_{61} \leq 1;$$

$$w_{43} + w_{21} + w_{31} + w_{11} + w_{51} + w_{61} \leq 1;$$

$$w_{53} + w_{21} + w_{31} + w_{41} + w11 + w_{61} \leq 1;$$

$$w_{63} + w_{21} + w_{31} + w_{41} + w_{51} + w_{11} \leq 1;$$

$$w_{12} + w_{22} + w_{32} + w_{42} + w_{52} + w_{62} = 1;$$

$$w_{11} + w_{13} + w_{21} + w_{23} + w_{31} + w_{33} + w_{41} + w_{43} + w_{51} + w_{53} + w_{61} + w_{63} = 2;$$

$$w_{13} \geq w_{12};$$

$$w_{12} \geq w_{11};$$

$$w_{11} > 0;$$

$$w_{23} \geq w_{22};$$

$$w_{22} \geq w_{21};$$

$$w_{21} > 0;$$

$$w_{33} \geq w_{32};$$

$$w_{32} \geq w_{31};$$

$$w_{31} > 0;$$

$$w_{43} \geq w_{42};$$

$$w_{42} \geq w_{41};$$

$$w_{41} > 0;$$

$$w_{53} \geq w_{52};$$

$$w_{52} \geq w_{51};$$

$$w_{51} > 0;$$

$$w_{63} \geq w_{62};$$

$$w_{62} \geq w_{61};$$

$$w_{61} > 0;$$

References

Jafarnejad, A., Soufi, M., & Bayati, A. (2014). Prioritizing critical barriers of computerized maintenance management system (CMMS) by fuzzy multi attribute decision making (F-MADM) (using LFPP). *Kuwait Chapter of Arabian Journal of Business and Management Review, 4*(3), 11–27.

Safari, H., & Soufi, M. (2014). Select a hypermarket location based on fuzzy multi criteria decision making (F-MCDM) techniques (hybrid of F-DELPHI, F-AHP, F-LLSM and F-PROMETHEE). *Kuwait Chapter of Arabian Journal of Business and Management Review, 4*(1), 76–95.

Safari, H., & Soufi, M. (2015). Prioritize barriers of e-factories in Iran's industries with hybrid multi criteria decision making (MCDM) techniques; (AHP-LLSM). *Indian Journal of Fundamental and Applied Life Sciences, 5*, 1329–1344.

Chapter 4
Distance-Based Techniques in MCDM

Learning Objectives
Upon completion of this chapter, the reader will be able to:

1. **Classify** the main families of distance-based multi-criteria decision-making (MCDM) techniques, including TOPSIS, VIKOR, ELECTRE, and PROMETHEE, and differentiate their underlying principles and application contexts.
2. **Apply** the TOPSIS (technique for order of preference by similarity to ideal solution) methodology to rank alternatives by calculating their relative closeness to ideal and anti-ideal solutions.
3. **Implement** the VIKOR (VlseKriterijumska Optimizacija I Kompromisno Resenje) method to identify compromise solutions by balancing group utility and individual regret.
4. **Analyze** the foundational concepts of the European School or "outranking" approach, including pseudo-criteria, preference thresholds, and pairwise relations, as exemplified by the ELECTRE family of methods.
5. **Utilize** the PROMETHEE (preference ranking organization method for enrichment evaluations) technique, along with its different versions (I, II, III), to establish partial or complete rankings based on positive and negative outranking flows.
6. **Evaluate** the strengths, limitations, and appropriate application scenarios of each technique by interpreting results from comprehensive case studies in areas such as supplier selection and project prioritization.

A. Amirteimoori, M. Soufi, *Industrial Decision Analysis*, Multiple Criteria Decision Making, https://doi.org/10.1007/978-3-032-24299-0_4

4.1 Introduction

Distance-based techniques in multi-criteria decision-making (MCDM) refer to methods that utilize the calculation of distances between alternatives and a reference point (such as an ideal or anti-ideal point) to rank and select the best alternative. These techniques are commonly used to evaluate and compare alternatives based on multiple criteria. Some of the most important distance-based techniques include:

1. TOPSIS (Technique for Order of Preference by Similarity to Ideal Solution)

- **Description**: This method calculates the distance of each alternative from two reference points: the "positive ideal solution" (best possible scenario) and the "negative ideal solution" (worst possible scenario).
- **Steps**:

  a. Normalize the decision matrix.
  b. Determine the positive and negative ideal solutions.
  c. Calculate the distance of each alternative from these points.
  d. Compute the closeness coefficient to the ideal solution.
  e. Rank the alternatives based on the closeness coefficient.

2. VIKOR (VIseKriterijumska Optimizacija I Kompromisno Resenje)

- **Description**: This method is based on the concept of "group utility" and "individual regret" and seeks to find a compromise solution.
- **Steps**:

  a. Normalize the decision matrix.
  b. Calculate the utility (S) and regret (R) measures for each alternative.
  c. Compute the VIKOR index (Q) for each alternative.
  d. Rank the alternatives based on the Q, S, and R values.

3. ELECTRE (Elimination and Choice Translating Reality)

- **Description**: This method uses pairwise comparisons of alternatives to establish outranking relationships and operates based on relative distances between alternatives.
- **Steps**:

  a. Construct concordance and discordance matrices.
  b. Calculate outranking indices.
  c. Eliminate nondominant alternatives.
  d. Select the best alternative.

4. SAW (Simple Additive Weighting)

- **Description**: This is the simplest distance-based technique, where each alternative is evaluated based on the weighted sum of normalized values.
- **Steps**:

a. Normalize the decision matrix.
b. Apply weights to the criteria.
c. Calculate the overall score for each alternative.
d. Rank the alternatives.

5. COPRAS (Complex Proportional Assessment)

- **Description**: This method calculates the relative utility and cost of each alternative.
- **Steps**:

a. Normalize the decision matrix.
b. Calculate the relative utility and cost measures.
c. Compute the COPRAS index for each alternative.
d. Rank the alternatives.

6. GRA (Grey Relational Analysis)

- **Description**: This method uses grey system theory and calculates the degree of similarity between alternatives and a reference point.
- **Steps**:

a. Normalize the decision matrix.
b. Calculate grey relational coefficients.
c. Compute the grey relational grade.
d. Rank the alternatives.

7. MOORA (Multi-objective Optimization on the Basis of Ratio Analysis)

- **Description**: This method calculates normalized ratios and combines them with criterion weights.
- **Steps**:

a. Normalize the decision matrix.
b. Calculate the overall score for each alternative.
c. Rank the alternatives.

These techniques assist decision-makers in selecting the best alternative by considering the relative distances between alternatives and reference points. The choice of the appropriate method depends on the nature of the problem, the number of criteria, and the preferences of the decision-maker.

4.2 Norms in Multi-criteria Decision-Making (MCDM)

In **multi-criteria decision-making (MCDM)** methods, **norms** are mathematical tools used to measure distances or differences between values. Norms are generally employed to calculate the distance between two points or two vectors in a multidimensional space and play a significant role in methods such as **TOPSIS**, **VIKOR**, and other MCDM techniques. In these methods, norms are typically used to compute the distance of alternatives from the positive ideal solution and the negative ideal solution.

4.2.1 *Definition of a Norm*

A norm is a mathematical function that assigns a non-negative real number to each vector in a vector space. This function has three key properties:

a. **Non-negativity**: $\|x\| \geq 0$, and $\|x\| = 0$ if and only if $x = 0$.
b. **Homogeneity**: $\|\alpha x\| = |\alpha| . \|x\|$for any scalar α.
c. **Triangle inequality**: $\|x + y\| \leq \|x\| + \|y\|$.

4.2.2 *Types of Norms*

In MCDM methods, various norms are used, each with its own specific characteristics. Some of the most important norms include:

a. **Euclidean Norm**

- This norm calculates the Euclidean distance between two points in a multidimensional space.
- The formula for the Euclidean norm of a vector $x = (x_1, x_2, \ldots, x_n)$ is:

$$\|x\|_2 = \sqrt{\sum_{i=1}^{n} x_i^2}.$$

- In methods such as **TOPSIS**, the Euclidean norm is used to compute the distance of alternatives from the positive and negative ideal solutions.

b. **Manhattan Norm**

- This norm calculates the sum of absolute differences between values.
- The formula for the Manhattan norm of a vector $x = (x_1, x_2, \ldots, x_n)$ is:

$$\|x\|_1 = \sum_{i=1}^{n} |x_i|$$

- This norm is used in methods such as **VIKOR**.

c. **Chebyshev Norm**

- This norm calculates the maximum difference between values.
- The formula for the Chebyshev norm of a vector $x = (x_1, x_2, \ldots, x_n)$ is: $\|x\|_\infty = \max \mid x_i \mid$.
- This norm is used in methods that emphasize the maximum difference.

d. **p-Norm (General Norm)**

- This norm generalizes the Euclidean and Manhattan norms and is defined as:

$$\|x\|_p = \sqrt[p]{\sum_{i=1}^{n} |x_i|^p}.$$

- For $p = 1$, it yields the Manhattan norm, and for $p = 2$, it yields the Euclidean norm.

e. **Weighted Norm**

- In this norm, each element of the vector is assigned a weight to reflect its importance in the distance calculation.
- The formula for the weighted norm of a vector $x = (x_1, x_2, \ldots, x_n)$with a weight vector $w = (w_1, w_2, \ldots, w_n)$is:

$$\|x\|_w = \sqrt{\sum_{i=1}^{n} w_i x_i^2}.$$

- This norm is used in methods such as **TOPSIS** to compute weighted distances.

4.2.3 Applications of Norms in MCDM Methods

- **TOPSIS**: Uses the Euclidean norm (often weighted) to calculate the distance of alternatives from the positive and negative ideal solutions
- **VIKOR**: Uses the Manhattan norm to compute distances
- **SAW (simple additive weighting)**: Uses the weighted norm to combine normalized values
- **ELECTRE**: Uses various norms to compute distances and establish outranking relationships

4.2.4 Selecting the Appropriate Norm

The choice of an appropriate norm depends on the nature of the problem and the type of data. For example:

- If the focus is on the **sum of differences**, the Manhattan norm is used.
- If the focus is on the **maximum difference**, the Chebyshev norm is used.
- If the focus is on the **geometric distance**, the Euclidean norm is used.

4.2.5 *Conclusion*

Norms are powerful mathematical tools used in MCDM methods to calculate distances and differences. The choice of an appropriate norm depends on the nature of the problem and the decision-making objectives. Commonly used norms include the Euclidean norm, Manhattan norm, Chebyshev norm, and weighted norm.

4.3 TOPSIS (Technique for Order of Preference by Similarity to Ideal Solution)

The TOPSIS method is one of the multi-attribute decision-making (MADM) techniques used for ranking alternatives. This method employs two key concepts: the "ideal solution" and "similarity to the ideal solution." As the name suggests, the ideal solution represents a solution that is optimal in every aspect, though it typically does not exist in practice, and efforts are made to approximate it. To measure the similarity of a design (or alternative) to the ideal and anti-ideal solutions, the distance of that design (or alternative) from both the ideal and anti-ideal solutions is calculated. The alternatives are then evaluated and ranked based on the ratio of the distance from the anti-ideal solution to the sum of the distances from the ideal and anti-ideal solutions. The term TOPSIS is derived from the acronym for **technique for order of preference by similarity to ideal solution**.

TOPSIS views an MADM problem with m alternatives as a geometric system consisting of m points in an n-dimensional space. The TOPSIS method is based on the concept that the chosen alternative should have the shortest distance from the positive ideal solution and the longest distance from the negative-ideal solution. TOPSIS defines an index called the similarity to the positive ideal solution and the remoteness from the negative-ideal solution. The method then selects the alternative with the maximum similarity to the positive ideal solution.

If an alternative is similar to an ideal solution, it is assigned a higher ranking. The ideal solution is one that is optimal in every aspect, though it does not practically exist, and we strive to approximate it. Essentially, to measure the similarity of a design (or alternative) to the ideal and nonideal levels, we consider the distance of that design from both the ideal and nonideal solutions.

4.3.1 *Advantages of the TOPSIS Method*

- **Decision-Making in the Presence of Positive and Negative Criteria**: The TOPSIS method enables decision-making even when both positive and negative criteria are present (even simultaneously in a single problem). Positive criteria are those that represent benefits, such as product quality, while negative criteria represent drawbacks, such as task difficulty.

- **Ability to Evaluate a Significant Number of Criteria**: This method allows for the evaluation of a considerable number of criteria to determine the best alternative. In contrast, methods like AHP or ANP inherently have limitations in this regard.
- **Simplicity and Appropriate Speed**: The TOPSIS method is straightforward and operates at a suitable speed, making it highly effective for problems involving a large number of alternatives and criteria.
- **Ease of Quantifying Qualitative Criteria**: In TOPSIS, qualitative criteria can easily be quantified, enabling decision-making that incorporates both qualitative and quantitative criteria.
- **Quantitative Output and Precise Ranking**: The system's output is quantitative, and in addition to identifying the best alternative, the rankings of other alternatives are expressed numerically. This numerical value, known as "relative closeness," forms the strong foundation of this method.
- **Strong Mathematical Basis**: The TOPSIS method is grounded in robust mathematical principles. It deals with distances and selects the alternative that has the maximum distance from the worst solution and the minimum distance from the best solution. This mathematical foundation gives TOPSIS an advantage over other MADM methods.
- **Compensatory Method and Inclusion of All Weights**: TOPSIS has another advantage over some MADM methods: it is a compensatory method. This means that the weights of all alternatives and criteria are incorporated into the decision-making process, and no weight is overlooked.

These advantages make the TOPSIS method one of the most powerful and widely used techniques in multi-attribute decision-making (MADM).

4.3.2 *Mathematical Foundation of TOPSIS*

a. **Formulation of the Decision Matrix**

- The decision matrix is constructed with alternatives represented as rows and criteria as columns. Each element x_{ij} in the matrix denotes the performance of the i*i*th alternative with respect to the *j*th criterion.

b. **Normalization of the Decision Matrix**

- To facilitate the comparison of criteria measured in different units, the decision matrix is normalized. The normalization formula is expressed as:

$$r_{ij} = \frac{x_{ij}}{\sqrt{\sum_{i=1}^{m} x_{ij}^2}}$$

where r_{ij} represents the normalized value of x_{ij}.

c. **Application of Weights to Criteria**

- Each criterion is assigned a weight w_j, reflecting its relative importance. The weighted matrix v_{ij} is computed as follows:

$$v_{ij} = w_j \times r_{ij}.$$

d. **Determination of the Positive and Negative Ideal Solutions**

- The positive ideal solution $A+$ and the negative ideal solution $A-$ are defined as:

$$A^+ = \{(\max v_{ij}| j \in J), (\min v_{ij}| j \in J')\}$$
$$A^- = \{(\min v_{ij}| j \in J), (\max v_{ij}| j \in J')\}$$

where J represents the set of beneficial criteria (to be maximized) and J' represents the set of non-beneficial criteria (to be minimized).

e. **Calculation of Distances from the Ideal Solutions**

- The Euclidean distance of each alternative from A^+ and A^- is calculated using the following formulas:

$$D_i^+ = \sum\nolimits_{i=1}^{n} \left(v_{ij} - A_j^+\right)^2$$
$$D_i^- = \sum\nolimits_{i=1}^{n} \left(v_{ij} - A_j^-\right)^2.$$

f. **Computation of Relative Closeness to the Ideal Solution**

- The relative closeness Ci of each alternative to the positive ideal solution is determined as:

$$C_i = \frac{D_i^-}{D_i^+ + D_i^-}.$$

- A higher value of Ci indicates a more favorable alternative.

4.3.3 Applications of TOPSIS

- **Supply Chain Management**: Selection of suppliers based on multiple criteria such as cost, quality, and delivery time

- **Project Management**: Evaluation and prioritization of projects based on criteria such as budget, timeline, and risk
- **Technology Selection**: Choosing the most suitable technology based on technical, economic, and operational criteria
- **Human Resource Management**: Assessing and selecting candidates or employees based on diverse performance metrics

4.3.4 Strengths of TOPSIS

- **Simplicity and Intuitiveness**: The methodology is straightforward and easy to comprehend, making it accessible to practitioners.
- **Flexibility in Weight Assignment**: The technique allows for the incorporation of varying weights to reflect the relative importance of different criteria.
- **Broad Applicability**: TOPSIS is versatile and can be applied across a wide range of decision-making contexts.

4.3.5 Limitations of TOPSIS

- **Sensitivity to Weight Assignment**: The results are highly dependent on the weights assigned to the criteria, which may introduce subjectivity.
- **Impact of Normalization Methods**: The choice of normalization technique can influence the outcomes, necessitating careful consideration.
- **Assumption of Criterion Independence**: TOPSIS assumes that criteria are independent, which may not hold true in all real-world scenarios.

4.3.6 Steps for Problem-Solving Using TOPSIS

- **Construct the Decision Matrix**: The initial step of this method is the formation of the decision matrix. The decision matrix in this method consists of a set of criteria and alternatives, structured as a matrix where the criteria are placed in the columns and the alternatives are placed in the rows. Each cell in the matrix represents the evaluation of an alternative with respect to a specific criterion. Once the decision matrix is formed, it must be completed using expert opinions. This process is carried out using Likert scales, Saaty scales, or real numbers. In cases where the criterion is quantitative, such as cost or production rate, the actual numerical value for each alternative is used. However, for qualitative criteria where numerical values are not inherently meaningful, scales such as 1–9 or 1–5 are employed.

- **Normalize the Decision Matrix**: Normalize the matrix to ensure comparability across criteria. In the TOPSIS method, normalization (or dimensionless scaling) is performed using the norm method. This process is carried out by dividing each entry of the decision matrix by the square root of the sum of the squares of the entries in the corresponding column (criterion). In other words, for each element x_{ij} in the decision matrix, the normalized value r_{ij} is calculated as follows:

$$r_{ij} = \frac{x_{ij}}{\sqrt{\sum_{i=1}^{m} x_{ij}^2}}.$$

In this step, the decision matrix is transformed into a **dimensionless matrix**, meaning that its values are unitless and comparable. This stage is essential to eliminate the influence of different measurement scales of the criteria and to establish a common basis for comparing alternatives.

- **Apply Criteria Weights**: Assign weights to each criterion and compute the weighted matrix. In this step, the weights of the criteria, which are obtained through other methods, must be multiplied by the normalized matrix to yield the weighted matrix. (The TOPSIS method alone is not capable of calculating the weights of the criteria; therefore, weights must be determined using other methods such as AHP, entropy, etc., and provided as input to this method.)
- **Identify Ideal Solutions**: Determine the positive ideal solution $A+$ and the negative ideal solution $A-$. Here, the type of criteria must be determined. Criteria can either have a positive aspect or a negative aspect. Positive criteria are those whose increase leads to an improvement in the system, such as the quality of a product. This type of criterion is considered positive, and its ideal solution corresponds to the largest value in the column of that criterion, while the anti-ideal solution corresponds to the smallest value in the column. For negative criteria, the opposite applies.

  For criteria with a **positive impact**, the **positive ideal** is the **largest value** of that criterion.

  For criteria with a **positive impact**, the **negative ideal** is the **smallest value** of that criterion.

  For criteria with a **negative impact**, the **positive ideal** is the **smallest value** of that criterion.

  For criteria with a **negative impact**, the **negative ideal** is the **largest value** of that criterion.
- **Calculate Distances**: Compute the Euclidean distances of each alternative from $A+$ and $A-$.
- **Determine Relative Closeness**: Calculate the relative closeness Ci for each alternative.
- **Rank the Alternatives**: Rank the alternatives based on their relative closeness scores, with higher values indicating better alternatives.

These steps provide a systematic approach for decision-makers to evaluate and select the most suitable alternative based on a comprehensive analysis of multiple criteria.

Example 4.1

Suppose we want to evaluate four alternatives (A, B, C, D) based on three criteria (C_1, C_2, C_3). Criteria C_1 and C_2 are of the maximization type (higher values are better), while C_3 is of the minimization type (lower values are better). The weights for the criteria are 0.4, 0.3, and 0.3, respectively. The decision matrix is as follows:

Alternative	C_1	C_2	C_3
A	*7*	*9*	*9*
B	*8*	*7*	*6*
C	*9*	*6*	*7*
D	*6*	*8*	*8*
W_i	*0.3*	*0.3*	*0.4*

Steps to Solve Using TOPSIS

1. Construct the Decision Matrix

The decision matrix X is:

	Alternative	C_1	C_2	C_3
$X =$	*A*	*7*	*9*	*9*
	B	*8*	*7*	*6*
	C	*9*	*6*	*7*
	D	*6*	*8*	*8*
	W_i	*0.3*	*0.3*	*0.4*

2. Normalize the Decision Matrix

Normalization is performed using the formula:

$$r_{ij} = \frac{x_{ij}}{\sqrt{\sum_{i=1}^{m} x_{ij}^2}}.$$

Calculations for each column (criterion):

- Criterion C_1:

$$7^2+8^2+9^2+6^2=49+64+81+36=230 \qquad \sqrt{230}\approx 15.16$$

$$r_{11}=\frac{7}{15.16}\approx 0.46,\ r_{21}=\frac{8}{15.16}\approx 0.53,\ r_{31}=\frac{9}{15.16}\approx 0.59,\ r_{41}=\frac{6}{15.16}\approx 0.40$$

- Criterion C_2:

$$9^2+7^2+6^2+8^2=81+49+36+64=230 \qquad \sqrt{230}\approx 15.16$$

$$r_{12}=\frac{9}{15.16}\approx 0.59,\ r_{22}=\frac{7}{15.16}\approx 0.46,\ r_{32}=\frac{6}{15.16}\approx 0.40,\ r_{42}=\frac{8}{15.16}\approx 0.53$$

- Criterion C_3:

$$9^2+6^2+7^2+8^2=81+36+49+64=230 \qquad \sqrt{230}\approx 15.16$$

$$r_{13}=\frac{9}{15.16}\approx 0.59,\ r_{23}=\frac{6}{15.16}\approx 0.40,\ r_{33}=\frac{7}{15.16}\approx 0.46,\ r_{43}=\frac{8}{15.16}\approx 0.53$$

The normalized matrix R is:

	Alternative	C_1	C_2	C_3
$R=$	*A*	*0.46*	*0.59*	*0.59*
	B	*0.53*	*0.46*	*0.40*
	C	*0.59*	*0.40*	*0.46*
	D	*0.40*	*0.53*	*0.53*
	W_i	*0.3*	*0.3*	*0.4*

3. Apply Weights to the Criteria
The weighted matrix V is calculated as: $v_{ij} = w_{ij} \times r_{ij}$.

	Alternative	C_1	C_2	C_3
$V=$	*A*	*0.138*	*0.177*	*0.236*
	B	*0.159*	*0.138*	*0.160*
	C	*0.177*	*0.120*	*0.184*
	D	*0.120*	*0.159*	*0.212*

4. Determine the Positive Ideal Solution (A^+) and Negative Ideal Solution (A^-)

Positive Ideal Solution (PIS): Maximum values for maximization criteria (C_1, C_2) and minimum values for minimization criteria (C_3)
Negative Ideal Solution (NIS): Minimum values for maximization criteria (C_1, C_2) and maximum values for minimization criteria (C_3)

Alternative	C_1	C_2	C_3
A	*0.138*	*0.177*	*0.236*
B	*0.159*	*0.138*	*0.160*
C	*0.177*	*0.120*	*0.184*
D	*0.120*	*0.159*	*0.212*
A^+	*0.177*	*0.177*	*0.160*
A^-	*0.120*	*0.120*	*0.236*

5. Calculate the Euclidean Distance from A⁺ and A⁻

The Euclidean distance of each alternative from A^+ and A^- is calculated.

For example, for alternative A:

$$D_{A^+} = \sqrt{(0.1385-0.1780)^2+(0.1780-0.1780)^2+(0.2373-0.1582)^2} \approx 0.089$$

$$D_{A^-} = \sqrt{(0.1385-0.1187)^2+(0.1780-0.1178)^2+(0.2373-0.2373)^2} \approx 0.062.$$

These calculations are repeated for the other alternatives:

Alternative	D_{A^+}	D_{A^-}
A	*0.089*	*0.062*
B	*0.045*	*0.089*
C	*0.062*	*0.089*
D	*0.089*	*0.062*

6. Calculate the Relative Closeness (C_i)

The relative closeness of each alternative to A^+ is calculated using the formula:

$$C_i = \frac{D_i^-}{D_i^+ + D_i^-}$$

$$C_A = \frac{0.062}{0.089+0.062} \approx 0.411$$

$$C_B = \frac{0.089}{0.045+0.089} \approx 0.666$$

$$C_C = \frac{0.089}{0.062+0.089} \approx 0.589$$

$$C_D = \frac{0.062}{0.089+0.062} \approx 0.411.$$

7. Rank the Alternatives

The alternatives are ranked based on their C_i values. The higher the C_i, the better the alternative.

Results

After performing the calculations, the alternatives are ranked as follows:

1. Alternative B (highest C_i)
2. Alternative C
3. Alternative A
4. Alternative D

This ranking indicates that Alternative B is the best choice.

4.4 Improved TOPSIS

This method is one of the modern multi-criteria decision-making (MCDM) techniques. It was proposed by **Niu et al.** (2017). They argued that in the traditional TOPSIS method, the distance of alternatives from the ideal and anti-ideal solutions is calculated using the Euclidean distance, which is a basic measure. Therefore, they suggested integrating two additional distance measures (the **Hamming distance** and the **grey relational distance**) into the traditional TOPSIS model. These three distances (Euclidean, Hamming, and grey relational) are then combined into a final distance measure using a multiplicative approach. According to Niu et al., the evaluation results obtained using the improved TOPSIS method are more efficient and convincing.

Improved TOPSIS is based on the concept that the selected alternative should simultaneously be far from the negative ideal solution and close to the positive ideal solution. TOPSIS typically identifies points that, as solutions, are simultaneously the farthest from the negative ideal and the closest to the positive ideal. In the traditional TOPSIS method, the Euclidean distance formula is used to calculate distances. However, if there is correlation among the criteria, the **Mahalanobis distance formula**[1] is employed for greater accuracy, and this approach is referred to as **improved TOPSIS**.

The steps for solving a problem in improved TOPSIS, up to the calculation of the weighted normalized matrix and the subsequent determination of the positive and negative ideal solutions, are the same as in the original TOPSIS method. The improved TOPSIS differs from the original method starting from the step where the distances of each matrix element from the ideal solutions are calculated. At this stage, as mentioned earlier, the calculations are performed using three distances and their combination, as described below:

4.4.1 Calculate Distances Using Multiple Measures

The distances from A+ and A− are calculated using three distance measures: Euclidean, Hamming, and grey distances.

(a) **Euclidean Distance**

The Euclidean distance for alternative i*i* from A+ and A− is:

[1] The **Mahalanobis distance** is a statistical distance measure used to calculate the distance between a point and a distribution or between two points in a multidimensional space. Unlike the Euclidean distance, which only measures the geometric distance between points, the Mahalanobis distance takes into account the correlation between variables and their different scales. This method is particularly useful when data exhibits correlation or varying variances.

$$D_i^+(E) = \sqrt{\sum_{j=1}^{n} \left(v_{ij} - v_j^+\right)^2}$$

$$D_i^-(E) = \sqrt{\sum_{j=1}^{n} \left(v_{ij} - v_j^-\right)^2}.$$

(b) **Hamming Distance**

The Hamming distance for alternative i*i* from A+ and A− is:

$$D_i^+(H) = \sum_{j=1}^{n} \left|v_{ij} - v_j^+\right|$$

$$D_i^+(H) = \sum_{j=1}^{n} \left|v_{ij} - v_j^-\right|.$$

(c) **Grey Distance**

The grey distance for alternative i*i* from A+ and A− is:

$$D_i^+(G) = \frac{1}{n} \sum_{j=1}^{n} \frac{\left|v_{ij} - v_j^+\right|}{\left|v_{ij} + v_j^+\right|}$$

$$D_i^-(G) = \frac{1}{n} \sum_{j=1}^{n} \frac{\left|v_{ij} - v_j^-\right|}{\left|v_{ij} + v_j^-\right|}.$$

4.4.2 Combine Distances

The combined distances D_i^+ and D_i^- are calculated using a weighted sum of the three distance measures:

$$D_i^+ = \alpha.D_i^+(E) + \beta.D_i^+(H) + \gamma.D_i^+(G)$$
$$D_i^- = \alpha.D_i^-(E) + \beta.D_i^-(H) + \gamma.D_i^-(G)$$

α, β, γ: Weights for Euclidean, Hamming, and grey distances, respectively (with $\alpha + \beta + \gamma = 1$).

4.4.3 Calculate the Composite Score (S_i)

The composite score S_i for each alternative is calculated as:

$$S_i = \frac{D_i^-}{D_i^+ + D_i^-}$$

S_i: Composite score of alternative *i*.
$0 \leq S_i \leq 1$: A higher S_i indicates a better alternative.

4.4.4 Calculate the Overall Value (Z_i)

In the **improved TOPSIS** method, after calculating the relative closeness index (C_i) or the composite score (S_i), another criterion called the **overall value** (Z_i) can be used for the final ranking of alternatives. This criterion combines multiple indices (such as C_i and S_i) and helps the decision-maker evaluate the alternatives more accurately. Below, the calculation of Z_i and the ranking of alternatives based on it are explained.

The overall value (Z_i) for each alternative i*i* is calculated as a combination of the relative closeness index (C_i) and the composite score (S_i). The general formula for calculating Z_i is as follows:

$$Z_i = \lambda.C_i + (1 - \lambda).S_i$$

C_i: Relative closeness index of alternative i*i* (based on Euclidean distance).
S_i: Composite score of alternative *i* (based on the combination of Euclidean, Hamming, and grey distances).
λ: Weight assigned to C_i (a value between 0 and 1).
$(1 - \lambda)$: Weight assigned to S_i.

Steps for Calculating Z_i

- **Calculate C_i:**

The relative closeness index C_i is calculated as follows:

$$C_i = \frac{D_i^-}{D_i^+ + D_i^-}$$

D_i^+: Euclidean distance from the positive ideal solution (PIS or A^+).
D_i^- : Euclidean distance from the negative ideal solution (NIS or A^-).

- **Calculate S_i:**

  The composite score S_i is calculated as follows:

$$S_i = \frac{D_i^-}{D_i^+ + D_i^-}.$$

Here, D_i^+ and D_i^- are calculated by combining Euclidean, Hamming, and grey distances.

- **Calculate Z_i:**

  The overall value Z_i is calculated as a weighted combination of C_i and S_i:

$$Z_i = \lambda.C_i + (1 - \lambda).S_i.$$

The value of λ is determined by the decision-maker and represents the relative importance of C_i compared to S_i.

4.4.5 *Ranking Alternatives Based on* Z_i

Alternatives are ranked based on their Z_i values, from highest to lowest. An alternative with a higher Z_i value is ranked better. In general:

If $Z_i = 1$: Alternative i is completely close to the positive ideal solution (A^+).
If $Z_i = 0$: Alternative i is completely close to the negative ideal solution (A^-).

4.4.6 *Advantages of Using* Z_i

- **Combination of Multiple Criteria**: Z_i combines two indices, C_i and S_i, providing more information to the decision-maker.
- **Flexibility**: By adjusting the value of λ, the relative importance of C_i and S_i can be controlled.
- **Higher Accuracy**: The use of Z_i improves the accuracy of alternative ranking.

In the **improved TOPSIS** method, the overall value (Z_i) is used as a composite criterion for ranking alternatives. By combining the relative closeness index (C_i) and the composite score (S_i), this method enhances the accuracy and flexibility of the decision-making process, helping the decision-maker select the best alternative with greater confidence.

4.5 Extended TOPSIS

Extended TOPSIS, introduced by Shih et al. (2007), is an advanced version of the classical TOPSIS method designed to address multi-criteria decision-making (MCDM) problems by considering dependencies between criteria and uncertainty in data. This method introduces improvements in the calculations and structure of TOPSIS to better handle more complex problems.

Note: This method (extended TOPSIS) is used when the decision matrix is constructed based on the opinions of multiple decision-makers. If the TOPSIS data is in the form of real, quantitative data or if the opinion of a single decision-maker is applied, this technique cannot be used. In such cases, it is recommended to use the **traditional TOPSIS** or **improved TOPSIS** methods instead.

4.5.1 Steps of the Extended TOPSIS Method

Step 1: Construct the Decision Matrix

The first step in all multi-criteria decision-making (MCDM) models is to construct the decision matrix. In the TOPSIS method, this matrix is represented as a matrix composed of the problem's criteria and alternatives. However, in the extended method, the decision matrix is provided to multiple experts, who express their opinions on the rating of each alternative with respect to each criterion using various linguistic scales.

Step 2: Normalize the Decision Matrix

In this step, the decision matrix of each respondent is normalized separately using the Euclidean norm.

Step 3: Weight the Normalized Matrix

In this step, the weights of the criteria are multiplied by the normalized matrix. The weights of the criteria should be calculated using other techniques, such as the entropy method, analytic hierarchy process (AHP), SWARA method, or other weighting methods in MADM.

Step 4: Determine the Positive (A+) and Negative (A-) Ideal Solutions

This step is similar to the classical TOPSIS method.

Step 5: Calculate the Distance of Alternatives from the Ideal Solutions

This step is also similar to the classical TOPSIS method, where the Euclidean distance is used to calculate the distances. However, the difference is that the distance calculation is performed separately for each respondent.

Step 6: Aggregate the Distances of Alternatives

In this step, the geometric mean is used to aggregate the distances of the alternatives from the ideal solutions, as calculated by each respondent. Essentially, this formula indicates that to obtain an aggregated distance, the geometric mean of the distances derived from all respondents' opinions should be taken:

$$\overline{S_l^+} = \left(\prod_{k=1}^{K} S_i^{k+}\right)^{\frac{1}{k}}$$

$$\overline{S_l^-} = \left(\prod_{k=1}^{K} S_i^{k-}\right)^{\frac{1}{k}}.$$

Step 7: Calculate the Score of Each Alternative

In this step, the score of each alternative is calculated using the following formula, and the alternatives are ranked accordingly. The score of each alternative is a value between 0 and 1, and the closer it is to 1, the better the alternative is:

$$\overline{C_l^+} = \frac{\overline{S_l^-}}{\overline{S_l^+} + \overline{S_l^-}}.$$

4.5.2 Key Features of Extended TOPSIS

Considering Interdependencies Between Criteria

- In the classical TOPSIS method, criteria are considered independent of each other. However, in practice, criteria may have correlations or dependencies.
- In extended TOPSIS, methods such as ANP (analytic network process) or DEMATEL are used to model the interdependencies between criteria.

Handling Uncertainty

- In many real-world problems, data may be imprecise, incomplete, or fuzzy.
- Extended TOPSIS employs fuzzy numbers or rough set theory to manage uncertainty in data.

Use of Dynamic Weights

- The weights of criteria in extended TOPSIS can be dynamically adjusted based on the problem's conditions.

Improved Distance Calculations

- This method utilizes more advanced distance measures such as fuzzy distance or grey distance.

4.6 Fuzzy TOPSIS

In many real-world scenarios, decision-making problems are accompanied by constraints, objectives, and outcomes that are not precisely defined. Following the introduction of fuzzy sets by Bellman and Zadeh in the field of multi-criteria decision-making (MCDM), many researchers turned their attention to decision-making in fuzzy environments. The integration of fuzzy set theory and multi-criteria decision-making led to the emergence of a new theory known as fuzzy multi-criteria decision-making (FMCDM). This theory provides models capable of handling incomplete and uncertain information. A crucial point is that in the process of evaluation, judgment, or decision-making, natural language is often used, where words lack precise and definitive meanings. Therefore, fuzzy numbers are required to express linguistic variables and quantitatively describe the subjective judgments of decision-makers. Commonly used fuzzy numbers include triangular, trapezoidal, and Gaussian fuzzy numbers.

The concept of linguistic variables, introduced by Zadeh in 1975, enables computations using words instead of numbers. As a result, linguistic terms defined by fuzzy sets are widely used in decision theory problems to model uncertain information. Reliable sources such as reference books, including Chen and Hwang (1992) titled *Fuzzy Multiple Attribute Decision-Making: Methods and Applications* (Lecture Notes in Economics and Mathematical Systems, Springer-Verlag), and review articles in the field of FMCDM are available. Additionally, in recent years, new methods have been explored in this domain.

Following the introduction of intuitionistic fuzzy sets by Atanassov (1986), where each element is characterized not only by a membership function (as in fuzzy sets) but also by a nonmembership function, interest in studying decision-making problems using intuitionistic fuzzy sets increased. As an extension of classical sets, fuzzy sets, and intuitionistic fuzzy sets, Smarandache (1999) first proposed the concept of neutrosophic sets. In the paper by Wang et al. (2005), set-theoretic operators on an instance of neutrosophic sets, called interval neutrosophic sets, were presented. Recently, neutrosophic sets have been applied in the field of multi-criteria decision-making (MCDM).

Furthermore, Torra and Narukawa (2009) and Torra (2010) introduced the concept of hesitant fuzzy sets, which provide a more flexible approach for decision-makers when expressing their opinions. Consequently, hesitant fuzzy sets have found extensive applications in multi-criteria decision-making problems.

4.7 Case Study

4.7.1 Case Study 1

Assessment of Internal and External Consistency of Intuitive Decision-Making Methods in Comparison with Three Multi-criteria Decision-Making Methods (Safari et al., 2024)

One of the important questions that has not yet been adequately answered is which multi-criteria decision-making (MCDM) method is more suitable for decision-making. Why have alternative methods not been used? To address this question, in this study, which was conducted for prioritizing projects in Iran's oil industry, the most appropriate method was selected using three statistical correlation methods.

The process was as follows: First, project prioritization was performed using an intuitive method (Fuzzy Delphi) and three multi-criteria decision-making methods (fuzzy TOPSIS, fuzzy PROMETHEE II, and fuzzy ELECTRE III). Then, the internal and external consistency of the prioritizations was examined using Kendall's coefficient of concordance, Spearman's correlation coefficient, and Pearson's correlation coefficient. The method with the highest correlation was considered the more suitable method.

Here, focusing on the topic of fuzzy TOPSIS, only the prioritization steps of this method are presented. A decision matrix was formed with eight criteria and six alternatives. Ten experts from ten affiliated companies of the National Iranian Oil Industry were asked to indicate the importance level of the criteria relative to each alternative on a ten-point scale in a questionnaire.

The results are presented in the tables below. Based on this, linguistic variables were converted into triangular fuzzy numbers, and the steps of the fuzzy TOPSIS process were followed, as shown in the subsequent tables.

In this section, the evaluations of all ten experts from ten companies are presented. To better organize the tables, the criteria are represented by abbreviations, and the projects are denoted by numbers.

Definition of Abbreviations and Project Numbers
Criteria

1. **C**: Cost
2. **EP**: Environmental protection
3. **DS**: Development speed
4. **FS**: Feasibility study
5. **IA**: Infrastructure access
6. **HR**: Human resources
7. **DB**: Distance from border
8. **PC**: Political considerations

Projects

1. **P1**: Exploration projects package

2. **P2**: Field development projects package
3. **P3**: Maintenance and production increase projects package
4. **P4**: Golshan and Ferdowsi oil fields package
5. **P5**: South oil-rich fields package (Soussangerd, Sohrab, Arvand, etc.)
6. **P6**: Western Iran oil fields package

Linguistic Variables

VL: Very low
MVL: Moderately very low
L: Low
M: Medium
MH: Moderately high
H: High
MVH: Moderately very high
VH: Very high

Note: In this study, in addition to the indicators, a weight has been assigned to each company.

Step 1: Gathering expert opinions through the TOPSIS questionnaire (Tables 4.1, 4.2, 4.3, 4.4, 4.5, 4.6, 4.7, 4.8, 4.9 and 4.10)

Company 1 (weight = 8)
Company 2 (weight = 10)
Company 3 (weight = 4)
Company 4 (weight = 2)
Company 5 (weight = 5)
Company 6 (weight = 3)
Company 7 (weight = 4)
Company 8 (weight = 6)
Company 9 (weight = 7)
Company 10 (weight = 10)

Step 2: In this stage, the opinions of the experts should be combined using the geometric mean. However, since the objective of the study is to examine the correlation of the experts' opinions in prioritizing the options, the prioritization of the options has been conducted separately for each expert. For this reason, the opinions of the ten experts were not combined. Consequently, the linguistic variables of each expert are converted into triangular fuzzy numbers.

Here, an example of converting the linguistic variables of the first expert into fuzzy numbers is provided. The same procedure is applied to the other individuals as well (Tables 4.11, 4.12 and 4.13).

The same process is applied to each of the ten experts. The positive and negative distances, the closeness ratio, and the ranking of the options according to each expert are presented in Table 4.14.

As mentioned, given the objective of the study, the prioritization was calculated separately based on the opinion of each expert. If needed, it is possible to use aggregation strategies in MCDM to combine these opinions and achieve an integrated ranking.

Table 4.1 Expert opinion of Company 1 regarding the importance of criteria for projects

Project	C	EP	DS	FS	IA	HR	DB	PC
P1	L	VL	L	M	M	H	H	VL
P2	ML	MH	VL	L	L	M	ML	L
P3	H	VH	H	M	M	H	VH	H
P4	H	M	MH	L	VL	L	M	M
P5	M	MH	M	L	VL	M	ML	H
P6	H	ML	H	L	VL	MH	ML	VH
Wj	L	M	MH	H	VH	ML	MH	H

Table 4.2 Expert opinion of Company 2 regarding the importance of criteria for projects

Project	C	EP	DS	FS	IA	HR	DB	PC
P1	H	L	L	ML	M	M	VH	M
P2	ML	VL	VL	L	H	ML	MH	L
P3	H	VH	H	M	M	H	M	VH
P4	VH	M	M	H	H	ML	MH	VL
P5	M	MH	M	MH	ML	MH	L	VH
P6	MH	ML	H	L	M	L	MH	VL
Wj	L	ML	M	ML	VH	M	MH	H

Table 4.3 Expert opinion of Company 3 regarding the importance of criteria for projects

Project	C	EP	DS	FS	IA	HR	DB	PC
P1	H	H	VL	H	MH	H	L	H
P2	M	ML	L	L	L	ML	H	VL
P3	H	VH	H	M	MH	H	M	MH
P4	L	M	M	ML	ML	VL	M	M
P5	M	ML	H	L	VL	M	ML	M
P6	MH	ML	VH	H	L	ML	ML	ML
Wj	ML	MH	H	VH	L	VL	M	M

Table 4.4 Expert opinion of Company 4 regarding the importance of criteria for projects

Project	C	EP	DS	FS	IA	HR	DB	PC
P1	H	L	H	ML	M	VL	H	MH
P2	ML	H	VL	L	H	L	L	L
P3	H	M	MH	M	M	H	M	MH
P4	VL	M	M	VH	MH	M	ML	ML
P5	M	ML	M	MH	VH	H	L	VL
P6	ML	ML	ML	L	H	VH	H	L
Wj	VL	M	M	L	VL	H	VH	L

Some of these strategies include:

- **Copeland method**: Aggregates results based on pairwise comparisons of alternatives
- **Borda method**: Combines rankings by assigning scores based on the position of each alternative in individual rankings
- **Average rank method**: Averages the ranks of each alternative across different methods to determine the final ranking

Table 4.5 Expert opinion of Company 5 regarding the importance of criteria for projects

Project	C	EP	DS	FS	IA	HR	DB	PC
P1	H	L	M	VH	M	L	VL	L
P2	ML	MH	ML	MH	L	ML	MH	VL
P3	VH	MH	H	M	VH	H	VH	H
P4	MH	L	ML	MH	VL	H	M	MH
P5	M	H	MH	L	VH	M	MH	M
P6	L	H	L	MH	VL	H	ML	H
Wj	VL	L	M	MH	H	L	M	MH

Table 4.6 Expert opinion of Company 6 regarding the importance of criteria for projects

Project	C	EP	DS	FS	IA	HR	DB	PC
P1	M	H	H	VL	M	L	VL	L
P2	L	M	ML	L	L	ML	MH	VL
P3	M	H	VH	H	VH	H	VH	H
P4	VL	L	M	M	VL	H	M	MH
P5	VL	M	ML	H	VH	M	MH	M
P6	VL	MH	ML	VH	VL	H	ML	H
Wj	VH	ML	MH	H	H	L	M	MH

Table 4.7 Expert opinion of Company 7 regarding the importance of criteria for projects

Project	C	EP	DS	FS	IA	HR	DB	PC
P1	H	L	H	H	VL	H	VL	L
P2	ML	MH	M	ML	L	L	MH	VL
P3	VH	MH	H	VH	H	M	VH	H
P4	MH	L	L	M	M	ML	M	MH
P5	M	H	M	ML	H	L	MH	M
P6	L	H	MH	ML	VH	H	ML	H
Wj	VL	L	ML	MH	H	VH	M	MH

Table 4.8 Expert opinion of Company 8 regarding the importance of criteria for projects

Project	C	EP	DS	FS	IA	HR	DB	PC
P1	H	L	M	VH	L	H	ML	M
P2	ML	MH	ML	MH	H	VL	L	H
P3	VH	MH	H	M	M	MH	M	M
P4	MH	L	ML	MH	M	M	VH	MH
P5	M	H	MH	L	ML	M	MH	VH
P6	L	H	L	MH	ML	ML	L	H
Wj	VL	L	M	MH	M	M	L	VL

- **Weighted average rank method**: Similar to the average rank method but assigns weights to each method based on its importance
- **Weighted linear combination**: Combines quantitative results from different methods using predefined weights
- **Group consensus method**: Seeks to achieve a consensus among experts using techniques like Delphi or group AHP

Table 4.9 Expert opinion of Company 9 regarding the importance of criteria for projects

Project	C	EP	DS	FS	IA	HR	DB	PC
P1	H	MH	H	L	M	VH	M	L
P2	L	L	ML	H	ML	MH	L	ML
P3	M	MH	H	M	H	M	VH	H
P4	ML	ML	VL	M	ML	MH	VL	H
P5	L	VL	M	ML	MH	L	VH	M
P6	H	L	ML	ML	L	MH	VL	H
Wj	VH	L	VL	M	M	MH	H	L

Table 4.10 Expert opinion of Company 10 regarding the importance of criteria for projects

Project	C	EP	DS	FS	IA	HR	DB	PC
P1	H	L	L	VL	H	L	H	ML
P2	ML	MH	ML	MH	ML	H	VL	L
P3	VH	MH	H	VH	H	M	MH	M
P4	MH	L	H	M	VL	M	M	VH
P5	M	H	M	MH	M	ML	M	MH
P6	L	H	H	ML	ML	ML	ML	L
Wj	VL	L	L	M	VL	M	M	L

- **Fuzzy integration method**: Combines results expressed as fuzzy numbers to handle uncertainty and ambiguity
- **Game theory-based integration**: Treats the results of different methods as players and combines them using game theory principles

These strategies enable the integration of diverse opinions and rankings into a unified and coherent decision-making framework.

4.7.2 *Case Study 2*

Identification and Ranking of Barriers to the Implementation of Green Supply Chain Management in Companies Affiliated with the Petrochemical Industries Using the Developed TOPSIS Method

Step 1: Introduction of abbreviations for decision variables and criteria (Tables 4.15 and 4.16)

Step 2: Formation of the decision-making matrix related to the barriers of green supply chain management (Table 4.17)

The weights of the criteria were calculated using the Shannon entropy method, and the results are presented in Table 4.18.

Step 3: Normalization of data for the purpose of dimensionless measurement units

Table 4.11 Conversion of the first expert's linguistic variables into triangular fuzzy numbers

Project	C			E P			DS			FS			IA			HR			DB			PC		
P1	0.1	0.2	0.3	0	0	0.2	0.1	0.2	0.3	0.4	0.5	0.6	0.4	0.5	0.6	0.7	0.8	0.9	0.7	0.8	0.9	0	0	0.2
P2	0.2	0.35	0.5	0.5	0.65	0.8	0	0	0.2	0.1	0.2	0.3	0.1	0.2	0.3	0.4	0.5	0.6	0.2	0.35	0.5	0.1	0.2	0.3
P3	0.7	0.8	0.9	0.8	0.9	1	0.7	0.8	0.9	0.4	0.5	0.6	0.4	0.5	0.6	0.7	0.8	0.9	0.8	0.9	1	0.7	0.8	0.9
P4	0.7	0.8	0.9	0.4	0.5	0.6	0.5	0.65	0.8	0.1	0.2	0.3	0	0	0.2	0.1	0.2	0.3	0.4	0.5	0.6	0.4	0.5	0.6
P5	0.4	0.5	0.6	0.5	0.65	0.8	0.4	0.5	0.6	0.1	0.2	0.3	0	0	0.2	0.4	0.5	0.6	0.2	0.35	0.5	0.7	0.8	0.9
P6	0.7	0.8	0.9	0.2	0.35	0.5	0.7	0.8	0.9	0.1	0.2	0.3	0	0	0.2	0.5	0.65	0.8	0.2	0.35	0.5	0.8	0.9	1
Wj	0.1	0.2	0.3	0.4	0.5	0.6	0.5	0.65	0.8	0.7	0.8	0.9	0.8	0.9	1	0.2	0.35	0.5	0.5	0.65	0.8	0.7	0.8	0.9

Table 4.12 Normalizing data using the Chebyshev norm

Normilize	C			E P			DS			FS			IA			HR			DB			PC		
P1	0.4	0.8	1.4	0.1	0.1	0.2	0.1	0.1	0.2	0.1	0.13	0.3	0.3	0.5	0.7	0.2	0.3	0.4	0.7	0.8	0.9	0	0	0.2
P2	0.1	0.4	0.8	0.1	0.00	0.1	0.0	0.0	0.1	0.1	0.8	0.2	0.6	0.8	1.0	0.1	0.2	0.3	0.2	0.3	0.5	0.1	0.2	0.3
P3	0.4	0.8	1.4	0.2	0.32	0.5	0.3	0.4	0.6	0.1	0.2	0.3	0.4	0.5	0.7	0.3	0.4	0.6	0.8	0.9	1	0.7	0.8	0.9
P4	0.4	0.9	1.5	0.1	0.18	0.3	0.2	0.3	0.4	0.2	0.3	0.5	0.6	0.8	1.0	0.1	0.2	0.3	0.4	0.5	0.6	0.4	0.5	0.6
P5	0.2	0.5	0.9	0.1	0.23	0.4	0.2	0.3	0.4	0.1	0.3	0.4	0.2	0.4	0.6	0.2	0.4	0.5	0.2	0.35	0.5	0.7	0.8	0.9
P6	0.3	0.7	1.2	0.1	0.18	0.3	0.3	0.4	0.6	0.1	0.1	0.2	0.4	0.5	0.7	0.1	0.1	0.2	0.2	0.35	0.5	0.8	0.9	1
Wj	0.4	0.8	1.4	0.1	0.07	0.2	0.1	0.1	0.2	0.1	0.1	0.3	0.4	0.5	0.7	0.2	0.3	0.4	0.7	0.8	0.9	0	0	0.2

Table 4.13 Weighted normalizing matrix and positive ideal solution and negative ideal solution

WN	C			EP			DS			FS			IA			HR			DB			PC		
P1	0.4	0.8	1.4	0.0	0.1	0.2	0.0	0.1	0.2	0.0	0.2	0.3	0.4	0.5	0.7	0.2	0.3	0.4	0.4	0.6	0.8	0.3	0.4	0.5
P2	0.1	0.4	0.8	0.0	0.0	0.1	0.0	0.0	0.1	0.0	0.1	0.2	0.6	0.8	1.0	0.1	0.2	0.3	0.3	0.4	0.6	0.1	0.2	0.3
P3	0.3	0.8	1.4	0.2	0.3	0.5	0.3	0.4	0.6	0.1	0.2	0.3	0.4	0.5	0.7	0.3	0.4	0.6	0.2	0.3	0.5	0.6	0.7	0.9
P4	0.4	0.9	1.5	0.1	0.2	0.3	0.2	0.3	0.4	0.2	0.3	0.5	0.6	0.8	1.0	0.1	0.2	0.3	0.3	0.4	0.6	0.0	0.0	0.2
P5	0.2	0.5	0.9	0.1	0.3	0.4	0.2	0.3	0.4	0.1	0.3	0.4	0.2	0.4	0.6	0.2	0.4	0.5	0.0	0.1	0.2	0.6	0.7	0.9
P6	0.3	0.7	1.2	0.1	0.2	0.3	0.3	0.4	0.6	0.0	0.1	0.2	0.4	0.5	0.7	0.1	0.1	0.2	0.3	0.4	0.6	0.0	0.0	0.2
I^+	0.4	0.9	1.5	0.2	0.3	0.5	0.3	0.4	0.6	0.2	0.3	0.5	0.6	0.8	1.0	0.3	0.4	0.6	0.4	0.6	0.8	0.6	0.7	0.9
I^-	0.1	0.4	0.8	0.0	0.0	0.1	0.0	0.0	0.1	0.0	0.1	0.2	0.2	0.4	0.6	0.1	0.1	0.2	0.0	0.1	0.2	0.0	0.0	0.2

Table 4.14 The positive and negative distances, the closeness ratio, and the ranking of the options according to each expert

d+	d−	CC	Rank
Expert 1			
2.220	1.155	0.342	4
1.979	0.088	0.043	6
0.005	2.198	0.998	1
0.752	1.274	0.629	3
1.401	0.638	0.313	5
0.667	1.587	0.704	2
Expert 2			
0.513	0.876	0.631	3
1.313	0.311	0.191	6
0.274	1.329	0.829	1
0.490	1.091	0.690	2
0.908	0.758	0.455	4
0.926	0.672	0.420	5
Expert 3			
0.463	3.078	0.869	2
2.104	1.287	0.379	5
0.114	3.033	0.964	1
2.894	0.158	0.052	6
1.793	1.480	0.452	4
0.811	2.593	0.762	3
Expert 4			
0.505	1.913	0.791	2
1.565	0.669	0.300	5
0.136	1.824	0.930	1
1.747	0.178	0.092	6
1.041	1.067	0.506	4
0.971	1.208	0.555	3
Expert 5			
0.669	1.541	0.697	2
1.608	0.550	0.255	5
0.059	2.223	0.974	1
0.913	1.297	0.587	3
1.047	1.159	0.525	4
1.943	0.368	0.159	6
Expert 6			
0.788	4711.145	1.000	2
2734.048	2209.485	0.447	3
0.005	4711.899	1.000	1
4711.477	0.256	0.000	6
4711.036	0.684	0.000	4
4711.381	0.566	0.000	5

(continued)

Table 4.14 (continued)

d+	d−	CC	Rank
Expert 7			
0.901	1.695	0.653	2
1.949	0.453	0.189	6
0.083	2.163	0.963	1
0.883	1.274	0.591	3
1.287	0.943	0.423	4
1.568	0.890	0.362	5
Expert 8			
0.364	1.554	0.810	2
1.245	0.573	0.315	5
0.107	1.658	0.939	1
0.550	1.189	0.684	3
1.064	0.760	0.417	4
1.650	0.150	0.083	6
Expert 9			
0.230	5.846	0.962	1
5.847	0.174	0.029	6
2.814	3.308	0.540	3
4.546	1.728	0.275	4
5.756	0.434	0.070	5
0.563	5.574	0.908	2
Expert 10			
0.506	1.406	0.736	2
1.203	0.493	0.291	5
0.043	1.629	0.975	1
0.531	1.105	0.675	3
0.903	0.720	0.444	4
1.604	0.074	0.044	6

Table 4.15 Introduction of abbreviations for criteria

Criteria	Abbreviation
Managerial measures	C1
Economic	C2
Social	C3
Environmental	C4
Governmental	C5
Supplier capabilities	C6
Organizational components	C7
Information technology	C8

Table 4.16 Introduction of abbreviations for decision variables

Decision variables (alternatives)	Abbreviation
Management commitment to GSCM	A1
Stability of internal company audits	A2
Lack of access to skilled human resources	A3
Capital required for GSCM implementation	A4
Emphasis on GSCM with ISO 14001 systems	A5
Pressure from various organizations and entities	A6
Environmental information disclosure	A7
Accountability to environmental authorities	A8
Environmental costs	A9
Achievement and development of green practices	A10
Role of government policies and guidelines	A11
Compliance with government regulations	A12
Respect for social rights of human resources	A13
Supplier commitment and engagement	A14
Volume of transactions in the past 2 years	A15
Supplier expertise and skills	A16
Organizational experience and expertise	A17
Environmental design requirements	A18
Process flexibility	A19
Integration of IT systems	A20
Company's technological level	A21
Adoption of new and advanced technologies	A22

In this step, the data are normalized to eliminate the influence of different measurement units, ensuring that all criteria are comparable on a dimensionless scale (Table 4.19).

Step 4: Creating the normalized weighted matrix (Table 4.20)

Step 5: Determining the positive and negative ideal solutions (Table 4.21)

Step 6: Calculating separate distances

In this step, the Mahalanobis distance is used to measure the distances of each alternative from the positive ideal solution and the negative ideal solution. This distance metric takes into account the correlations between variables, providing a more accurate measure of separation in multidimensional space compared to the Euclidean distance (Table 4.22).

Step 7: Calculating the relative closeness coefficient (Table 4.23)

Step 8: Prioritization of alternatives (Table 4.24)

In Table 4.25, the ranking of criteria and sub-criteria is presented in a comprehensive table.

Table 4.17 Decision-making matrix

Criteria	C1	C2	C3	C4	C5	C6	C7	C8
A1	2	1	3	2	4	3	2	2
A2	1	3	3	1	2	3	3	1
A3	4	1	4	2	4	2	3	3
A4	2	1	3	2	2	2	3	1
A5	2	4	1	4	2	4	3	2
A6	2	4	1	4	3	3	1	3
A7	4	4	4	4	1	2	3	4
A8	2	4	4	4	1	1	1	4
A9	1	3	3	3	2	4	3	1
A10	3	3	4	1	2	4	1	3
A11	4	4	2	2	3	3	1	1
A12	2	2	2	3	4	3	1	1
A13	4	2	4	1	1	3	1	4
A14	2	3	2	2	2	3	3	3
A15	4	2	1	2	4	4	1	2
A16	2	3	2	1	2	1	2	1
A17	3	4	2	1	1	4	3	1
A18	3	2	4	4	2	3	3	3
A19	3	4	4	2	2	4	2	4
A20	1	2	3	2	2	1	3	1
A21	2	1	4	3	2	3	2	4
A22	4	1	3	2	4	2	1	4

Table 4.18 Determination of criteria weights using the Shannon entropy method

Criteria	C1	C2	C3	C4	C5	C6	C7	C8
Impact	Positive	Positive	Positive	Positive	Positive	Positive	Positive	Positive
Weight (entropy)	0.107	0.136	0.1	0.135	0.122	0.09	0.13	0.181

4.8 The VIKOR Method

In multi-criteria decision-making models, ranking research alternatives is of great importance. One of the methods that address this issue is VIKOR. The VIKOR method originates from a Serbian term meaning "compromise solution and multi-criteria optimization," equivalent to the English phrase *multi-criteria optimization and compromise solution*. It was first introduced by Opricovic and Tzeng.

The VIKOR method is a multi-criteria decision-making approach designed to solve decision-making problems with conflicting criteria and different units of measurement. VIKOR focuses on ranking and selecting from a set of alternatives in problems with opposing criteria. The results obtained from the VIKOR method include a ranked list of compromise solutions, along with one or more optimal compromise solutions.

Table 4.19 Normalized matrix

Criteria	C1	C2	C3	C4	C5	C6	C7	C8
A1	0.153	0.074	0.21	0.164	0.331	0.214	0.187	0.158
A2	0.076	0.222	0.21	0.082	0.166	0.214	0.281	0.079
A3	0.306	0.074	0.279	0.164	0.331	0.143	0.281	0.236
A4	0.153	0.074	0.21	0.164	0.166	0.143	0.281	0.079
A5	0.153	0.296	0.07	0.329	0.166	0.286	0.281	0.158
A6	0.153	0.296	0.07	0.329	0.248	0.214	0.094	0.236
A7	0.306	0.296	0.279	0.329	0.083	0.143	0.281	0.315
A8	0.153	0.296	0.279	0.329	0.083	0.071	0.094	0.315
A9	0.076	0.222	0.21	0.247	0.166	0.286	0.281	0.079
A10	0.229	0.222	0.279	0.082	0.166	0.286	0.094	0.236
A11	0.306	0.296	0.14	0.164	0.248	0.214	0.094	0.079
A12	0.153	0.148	0.14	0.247	0.331	0.214	0.094	0.079
A13	0.306	0.148	0.279	0.082	0.083	0.214	0.094	0.315
A14	0.153	0.222	0.14	0.164	0.166	0.214	0.281	0.236
A15	0.306	0.148	0.07	0.164	0.331	0.286	0.094	0.158
A16	0.153	0.222	0.14	0.082	0.166	0.071	0.187	0.079
A17	0.229	0.296	0.14	0.082	0.083	0.286	0.281	0.079
A18	0.229	0.148	0.279	0.329	0.166	0.214	0.281	0.236
A19	0.229	0.296	0.279	0.164	0.166	0.286	0.187	0.315
A20	0.076	0.148	0.21	0.164	0.166	0.071	0.281	0.079
A21	0.153	0.074	0.279	0.247	0.166	0.214	0.187	0.315
A22	0.306	0.074	0.21	0.164	0.331	0.143	0.094	0.315

4.8.1 Steps of Problem-Solving Using the VIKOR Method

The steps for identifying criteria and alternatives, constructing the decision matrix, normalizing the data, forming the weighted normalized matrix, and determining the positive and negative ideal solutions are similar to those in the TOPSIS method. From this stage onward, the procedure differs:

- **Calculation of utility (S) and regret (R) values for each alternative:**
  The utility value (S) represents the relative distance of the ith alternative from the ideal solution, while the regret value (R) represents the maximum dissatisfaction of the ith alternative due to its distance from the ideal solution. These values are derived from the following formulas:

1. **Utility measure (S):**
   The utility measure S_i for the ith alternative is calculated as the weighted sum of the distances from the ideal solution:

Table 4.20 Waited normalized matrix

Criteria	C1	C2	C3	C4	C5	C6	C7	C8
A1	0.016	0.01	0.021	0.022	0.04	0.019	0.024	0.029
A2	0.008	0.03	0.021	0.011	0.02	0.019	0.037	0.014
A3	0.033	0.01	0.028	0.022	0.04	0.013	0.037	0.043
A4	0.016	0.01	0.021	0.022	0.02	0.013	0.037	0.014
A5	0.016	0.04	0.007	0.044	0.02	0.026	0.037	0.029
A6	0.016	0.04	0.007	0.044	0.03	0.019	0.012	0.043
A7	0.033	0.04	0.028	0.044	0.01	0.013	0.037	0.057
A8	0.016	0.04	0.028	0.044	0.01	0.006	0.012	0.057
A9	0.008	0.03	0.021	0.033	0.02	0.026	0.037	0.014
A10	0.025	0.03	0.028	0.011	0.02	0.026	0.012	0.043
A11	0.033	0.04	0.014	0.022	0.03	0.019	0.012	0.014
A12	0.016	0.02	0.014	0.033	0.04	0.019	0.012	0.014
A13	0.033	0.02	0.028	0.011	0.01	0.019	0.012	0.057
A14	0.016	0.03	0.014	0.022	0.02	0.019	0.037	0.043
A15	0.033	0.02	0.007	0.022	0.04	0.026	0.012	0.029
A16	0.016	0.03	0.014	0.011	0.02	0.006	0.024	0.014
A17	0.025	0.04	0.014	0.011	0.01	0.026	0.037	0.014
A18	0.025	0.02	0.028	0.044	0.02	0.019	0.037	0.043
A19	0.025	0.04	0.028	0.022	0.02	0.026	0.024	0.057
A20	0.008	0.02	0.021	0.022	0.02	0.006	0.037	0.014
A21	0.016	0.01	0.028	0.033	0.02	0.019	0.024	0.057
A22	0.033	0.01	0.021	0.022	0.04	0.013	0.012	0.057

Table 4.21 Positive and negative ideal solution

Criteria	Positive ideal solution	Negative ideal solution
C1	0.033	0.008
C2	0.040	0.010
C3	0.028	0.007
C4	0.044	0.011
C5	0.040	0.010
C6	0.026	0.006
C7	0.037	0.012
C8	0.057	0.014

$$S_i = \sum_{j=1}^{n} w_i \frac{f_j^+ - f_{ij}}{f_j^+ - f_j^-}$$

where:

w_j: Weight of the jth criterion
f_j^+: Positive ideal value for the jth criterion
f_j^-: Negative ideal value for the jth criterion
f_{ij}: Performance value of the ith alternative for the jth criterion

Table 4.22 Determining the Mahalanobis distance from the positive and negative ideal solutions

Alternatives	Positive (S_i^+)	Negative (S_i^-)
A1	0.052	0.043
A2	0.065	0.039
A3	0.042	0.060
A4	0.064	0.034
A5	0.044	0.058
A6	0.041	0.059
A7	0.033	0.075
A8	0.047	0.066
A9	0.056	0.047
A10	0.050	0.050
A11	0.057	0.048
A12	0.059	0.042
A13	0.055	0.056
A14	0.041	0.049
A15	0.052	0.049
A16	0.067	0.028
A17	0.064	0.048
A18	0.033	0.060
A19	0.033	0.065
A20	0.065	0.034
A21	0.044	0.057
A22	0.047	0.061

2. **Regret measure (R):**
   The regret measure R_i for the ith alternative is calculated as the maximum weighted distance from the ideal solution:

$$R_i = \max_j \left(w_i \frac{f_j^+ - f_{ij}}{f_j^+ - f_j^-} \right).$$

Explanation of Terms

Utility (S): Represents the overall weighted distance of an alternative from the ideal solution. A smaller S_i indicates a better alternative. Focuses on the overall performance of the alternative.

Regret (R): Represents the worst-case scenario for an alternative, i.e., the maximum dissatisfaction due to its distance from the ideal solution. A smaller R_i indicates a better alternative. Focuses on the worst-performing criterion for the alternative.

Table 4.23 Calculation of the relative closeness coefficient

Alternatives	Ci
A1	0.453
A2	0.375
A3	0.125
A4	0.347
A5	0.569
A6	0.590
A7	0.694
A8	0.584
A9	0.456
A10	0.500
A11	0.457
A12	0.416
A13	0.505
A14	0.544
A15	0.485
A16	0.295
A17	0.428
A18	0.154
A19	0.663
A20	0.343
A21	0.564
A22	0.565

Key Points
Balancing S and R: The VIKOR method uses these measures to find a compromise solution that balances group utility and individual regret.

- **Calculation of the VIKOR index (Q) for each alternative:**
  The VIKOR index (Q) for each alternative is calculated using the following formula:

$$Q_i = v\left(\frac{S_i - S^+}{S^- - S^+}\right) + (1 - v)\left(\frac{R_i - R^+}{R^- - R^+}\right)$$

where:

Q_i: VIKOR index for the ith alternative
v: Weight for the strategy of maximum group utility (usually set to 0.5)
S_i: Utility measure for the ith alternative
S^+: Maximum value of SS (worst utility measure)
S^-: Minimum value of SS (best utility measure)
R_i: Regret measure for the ith alternative
R^+: Maximum value of RR (worst regret measure)

Table 4.24 Ranking of alternatives

Alternatives	Rank
Management commitment to green supply chain management	14
Stability in evaluating and auditing internal company actions	17
Lack of access to skilled human resources	22
Capital required for implementing green supply chain management	18
Emphasis on supply chain management with ISO 14001 system	5
Pressure from various organizations and institutions	3
Environmental information disclosure	1
Accountability to environmental regulatory bodies	4
Environmental costs	13
Realization and development of green practices	10
Role of government policies and guidelines	12
Compliance with government laws and regulations	16
Respect for social rights of human resources	9
Supplier commitment and participation considering time and location	8
Volume of transactions in the past 2 years	11
Expertise and skills of the supplier	20
Organizational experience, knowledge, and expertise	15
Environmental design requirements	21
Process flexibility	2
Integration of information technology systems	19
Company's technological level	7
Adoption of new and advanced technologies	6

R^{-}: Minimum value of R*R* (best regret measure)

VIKOR Index (Q) Represents a compromise solution that balances group utility and individual regret. A smaller Q_i indicates a better alternative. The VIKOR index (Q) provides a ranking of alternatives based on their proximity to the ideal solution while considering both group utility and individual regret. The alternative with the smallest Q_i is considered the best compromise solution.

Weight (v) Reflects the decision-maker's preference for maximum group utility (v) versus minimum individual regret (1−v). Typically, v = 0.5 is used to balance both strategies.

Normalization The terms $\frac{S_i - S^+}{S^- - S^+}$ and $\frac{R_i - R^+}{R^- - R^+}$ normalize the utility and regret measures to a scale of 0–1.

Ranking Alternatives Based on S, Q, and R In the final step of the VIKOR technique, the alternatives are ranked in ascending order based on the values of Q, R, and S. The best alternative is the one that ranks highest in all three values (Q, R, and S). If such an alternative does not exist, the alternative with the smallest Q value is selected as the best option, provided that the following two conditions are met:

Table 4.25 Ranking of criteria and sub-criteria of green supply chain management barriers

Criteria	Rank of criteria	Sub-criteria	Rank of sub-criteria
Managerial actions	6	Management commitment to green supply chain management	14
		Stability in evaluating and auditing internal company actions	17
		Lack of access to skilled human resources	22
Economic	2	Capital required for implementing green supply chain management	18
Social	7	Emphasis on supply chain management with ISO 14001 system	5
		Pressure from various organizations and institutions	3
		Environmental information disclosure	1
		Accountability to environmental regulatory bodies	4
Environmental	3	Environmental costs	13
		Realization and development of green practices	10
Governmental	5	Role of government policies and guidelines	12
		Compliance with government laws and regulations	16
		Respect for social rights of human resources	9
Supplier capabilities	8	Supplier commitment and participation considering time and location	8
		Volume of transactions in the past 2 years	11
		Expertise and skills of the supplier	20
Organizational components	4	Organizational experience, knowledge, and expertise	15
		Environmental design requirements	21
		Process flexibility	2
Information technology	1	Integration of information technology systems	19
		Company's technological level	7
		Adoption of new and advanced technologies	6

Condition 1: Acceptable Advantage

If alternatives A_1 and A_2 are ranked first and second among mm alternatives, the following condition must be satisfied:

$$Q(A_2) - Q(A_1) \geq \frac{1}{m-1}$$

where:

$Q(A_1)$: VIKOR index of the first-ranked alternative.
$Q(A_2)$: VIKOR index of the second-ranked alternative.

m: Total number of alternatives. This condition ensures that the advantage of the first-ranked alternative over the second-ranked alternative is significant.

Condition 2: Acceptable Stability

The first-ranked alternative must also rank first in at least one of the S or R values. This ensures the stability of the solution.

Ranking: Alternatives are ranked based on Q, R, and S values.

Best Alternative: The alternative with the smallest Q value is selected as the best, provided the two conditions above are satisfied.

If Condition 1 is not met: Both alternatives A_1 and A_2 are considered the best options.

If Condition 2 is not met: Both alternatives A_1 and A_2 are selected as the top choices.

Explanation of Conditions

Condition 1 (Acceptable Advantage)

Ensures that the advantage of the first-ranked alternative (A_1) over the second-ranked alternative (A_2) is significant. If this condition is not met, both A_1 and A_2 are considered equally valid.

Condition 2 (Acceptable Stability)

Ensures that the first-ranked alternative (A_1) is stable and performs well in at least one of the R or S measures. If this condition is not met, both A_1 and A_2 are selected as the top alternatives.

Key Points

Compromise Solution: VIKOR aims to provide a compromise solution that balances group utility and individual regret.

Flexibility: If the conditions are not met, the method allows for multiple alternatives to be considered as the best options.

Determining the Parameter v in the VIKOR Method

One of the steps in the VIKOR method is determining the parameter v, which reflects the level of agreement among the decision-making groups. The parameter v balances the importance of group utility (majority opinion) and individual regret (minority opinion).

The value of v is determined as follows:

If there is very high agreement, $v > 0.5$: The decision-making process prioritizes the majority opinion, reflecting high group agreement.

If there is majority agreement, $v = 0.5$: The decision-making process equally balances group utility and individual regret, reflecting majority agreement.

If there is low agreement, $v < 0.5$: The decision-making process prioritizes individual opinions, reflecting low group agreement.

The parameter v represents the weight given to the strategy of maximum group utility versus minimum individual regret:

A larger v: Indicates greater emphasis on group opinions
A smaller v: Indicates greater emphasis on individual opinions

4.8.2 Differences Between the TOPSIS and VIKOR Methods

The TOPSIS technique, like the VIKOR method, aims to rank research alternatives. However, despite their similarities, these two methods have distinct differences, as outlined below:

1. **Approach to Ideal Solutions**
   TOPSIS: Considers the relative importance of distances from the positive and negative ideal solutions. As a result, the best solution in TOPSIS is not always the one closest to the positive ideal solution.
   VIKOR: Focuses on achieving a compromise solution by balancing group utility and individual regret, using the parameter v to reflect the level of agreement among decision-makers. This optimization makes VIKOR superior in decision-making scenarios.
2. **Optimization and Agreement**
   TOPSIS: Does not explicitly incorporate group agreement or compromise in its calculations
   VIKOR: Uses the parameter v to optimize decisions based on collective agreement, making it more effective for complex decision-making problems
3. **Purpose of Use**
   TOPSIS: More suitable when the goal is to rank alternatives
   VIKOR: More suitable when the goal is to select the best alternative, as it provides a compromise solution that balances conflicting criteria

Example 4.2
A website specializing in military weaponry aims to determine the best available fighter aircraft. To achieve this, they compared five different models of fighter aircraft and evaluated them based on six criteria: the number of missiles they can carry, weight, fuel consumption, maximum speed, operational range, and service life. The evaluations were conducted by experts, and the data is recorded in the table below. Using the VIKOR technique, identify the top fighter aircraft or aircrafts.

Criteria →		Number of missiles	Weight	Fuel consumption	Maximum speed	Operational range	Service life
Alternatives ↓		X_1^+	X_2^-	X_3^-	X_4^+	X_5^+	X_6^+
F–22 Raptor	A_1	12	19.4	6	2.5	2960	30
F/A–18 Hornet	A_i	7	18.8	5	2.0	3310	17
Sukhoi Su–27	A_r	8	20.7	9	2.1	2750	22
Typhoon Fr	A_v	8	20.7	6	1.5	2500	24
MiG–31	A_β	10	18.1	7	1.9	3060	42
w_j		0.14	0.11	0.24	0.18	0.18	0.15

Solution
Step 1: Normalize the decision matrix.

For benefit criteria $\left(X_1^+, X_4^+, X_5^+, X_6^+\right)$:

$$r_{ij} = \frac{x_{ij} - x_j^{\min}}{x_j^{\max} - x_j^{\min}}.$$

For cost criteria $\left(X_2^-, X_3^-\right)$:

$$r_{ij} = \frac{x_j^{\max} - x_{ij}}{x_j^{\max} - x_j^{\min}}.$$

Values of $x_j^{\max}$ and $x_j^{\min}$

Criteria	$x_j^{\max}$	$x_j^{\min}$
Number of missiles (X_1^+)	12	7
Weight (X_2^-)	18.1	20.7
Fuel consumption (X_3^-)	5	9
Maximum speed (X_4^+)	2.5	1.5
Operational range (X_5^+)	3310	2500
Service life (X_6^+)	42	17

Normalized matrix (r_{ij})

Aircraft	X_1^+	X_2^-	X_3^-	X_4^+	X_5^+	X_6^+
F–22 Raptor (A_1)	1.0	0.5	0.75	1.0	0.56	0.52
F/A–18 Hornet (A_2)	0.0	0.73	1.0	0.5	1.0	0.0
Sukhoi Su–27 (A_3)	0.2	0.0	0.0	0.6	0.31	0.2
Typhoon Fr (A_4)	0.2	0.0	0.75	0.0	0.0	0.28
MiG–31 (A_5)	0.6	1.0	0.5	0.4	0.69	1.0

Step 2: Weighted normalized matrix (v_{ij})

Multiply each normalized value by its corresponding weight:

$$v_{ij} = w_j \times r_{ij}.$$

Weighted normalized matrix (v_{ij}):

Aircraft	X_1^+	X_2^-	X_3^-	X_4^+	X_5^+	X_6^+
F–22 Raptor (A_1)	0.14	0.055	0.18	0.18	0.10	0.078
F/A–18 Hornet (A_2)	0.0	0.080	0.24	0.09	0.18	0.0
Sukhoi Su–27 (A_3)	0.028	0.0	0.0	0.108	0.056	0.03
Typhoon Fr (A_4)	0.028	0.0	0.18	0.0	0.0	0.042
MiG–31 (A_5)	0.084	0.11	0.12	0.072	0.124	0.15

Step 3: Determine best and worst values (f_j^+ and f_j^-).

For each criterion, extract the best and worst values from the weighted normalized matrix:

Criteria	f_j^+ (Best)	f_j^- (Worst)
Number of missiles (X_1^+)	0.14	0.0
Weight (X_2^-)	0.11	0.0
Fuel consumption (X_3^-)	0.24	0.0
Maximum speed (X_4^+)	0.18	0.0
Operational range (X_5^+)	0.18	0.0
Service life (X_6^+)	0.15	0.0

Step 4: Calculate S_i and R_i:

$$S_i = \sum_{j=1}^{n} w_i \frac{f_j^+ - f_{ij}}{f_j^+ - f_j^-}$$

$$R_i = \max_j \left(w_i \frac{f_j^+ - f_{ij}}{f_j^+ - f_j^-} \right).$$

Calculations for each aircraft

Aircraft	S_i	R_i
F–22 Raptor (A_1)	0.140	0.180
F/A–18 Hornet (A_2)	0.240	0.240
Sukhoi Su–27 (A_3)	0.028	0.108
Typhoon Fr (A_4)	0.028	0.180
MiG–31 (A_5)	0.084	0.150

Step 5: Calculate the VIKOR index (Q_i):

$$Q_i = v\left(\frac{S_i - S^+}{S^- - S^+}\right) + (1 - v)\left(\frac{R_i - R^+}{R^- - R^+}\right).$$

Assume v = 0.5.
Values of S^+, S^-, R^+, R^-:

$$S^+ = \min(S_i) = 0.028$$

$$S^- = \max(S_i) = 0.24$$

$$R^+ = \min(R_i) = 0.108$$

$$R^- = \max(R_i) = 0.24$$

Calculate Q_i:

Aircraft	Q_i
F–22 Raptor (A_1)	0.140
F/A–18 Hornet (A_2)	0.240
Sukhoi Su–27 (A_3)	0.028
Typhoon Fr (A_4)	0.028
MiG–31 (A_5)	0.084

Step 6: Rank the aircraft.

Based on the Q_i values, the aircraft are ranked as follows:

1. **Sukhoi Su–27 (A_3)** and **Typhoon Fr (A_4)** with $Q_i = 0.028$ (best)
2. **MiG–31 (A_5)** with $Q_i = 0.084$
3. **F–22 Raptor (A_1)** with $Q_i = 0.14$
4. **F/A–18 Hornet (A_2)** with $Q_i = 0.24$ (worst)

Conclusion

The best aircraft according to the VIKOR method are **Sukhoi Su–27** and **Typhoon Fr**.

4.9 Case Study

Prioritization of Green Supply Chain Suppliers Using an Integrated Fuzzy MCDM Approach (Homayounfar et al., 2018)

The increase in production levels of manufacturing systems, which is an unintended consequence of industrial development, has experienced significant growth in recent years. The negative impacts of this phenomenon have compelled companies to adopt strategies in the field of environmental management to maintain their competitive position. Among these strategies is the focus on green issues within the supply chain, which is largely influenced by the performance of their suppliers. Therefore, the evaluation and selection of green suppliers is considered a strategic decision for companies to remain competitive in the market and preserve their social standing.

Given the diversity of evaluation criteria, their quantitative or qualitative nature, and the fuzzy characteristics of qualitative criteria, this study proposes an integrated fuzzy MCDM approach based on fuzzy Delphi techniques, fuzzy analytic hierarchy process (FAHP), and fuzzy VIKOR for the evaluation and selection of green suppliers in an Iranian automotive company. Here, the calculations related to VIKOR are presented. In this study, 100 suppliers were evaluated based on seven criteria.

Step 1: Aggregating expert opinions using the geometric mean and forming the integrated decision matrix

First, the suppliers are evaluated by three experts (second-tier experts) based on the defined criteria. Subsequently, the integrated matrix of expert opinions is constructed, along with the weights assigned to each criterion (Table 4.26).

Step 2: Determining the best and worst values for each criterion across the alternatives

Next, the best and worst values for each criterion are identified based on the performance of the alternatives (suppliers). These values represent the optimal and suboptimal performance levels for each criterion within the evaluated set of suppliers (Table 4.27).

Based on the best and worst values for each criterion, the distances from the ideal solution and the regret values for each alternative are calculated. Then, the value of $\widetilde{Q}_i$ is calculated based on the best and worst fuzzy values for $\widetilde{S}_i$ and $\widetilde{R}_i$ which is used to perform the ranking (Table 4.28).

4.10 Superiority Techniques

The superiority approach, also known as the European school of multi-criteria decision-making, is one of the most important approaches in this field. It offers a greater diversity of techniques compared to other approaches and subfields of multi-criteria decision-making, and this diversity stems from the foundations of this approach. This approach is called the European school because it began in Europe with the ELECTRE I technique, which was introduced by Roy (1968). Additionally,

Table 4.26 Integrated matrix of supplier evaluations with respect to each criterion in the green supply chain

	C1			C2			C3			C4			C5			C6			C7		
A1	0.05	0.6	0.84	0.32	0.69	0.79	0.04	0.4	0.82	0.2	0.61	0.98	0.21	0.58	0.92	0.39	0.42	0.84	0.3	0.5	0.9
A2	0.32	0.62	0.96	0.27	0.53	0.94	0.27	0.62	0.85	0.14	0.65	0.91	0.04	0.6	0.84	0.21	0.42	0.96	0.2	0.5	0.8
A3	0.36	0.5	0.9	0.31	0.45	0.94	0.05	0.59	0.73	0.39	0.4	0.8	0.07	0.69	0.87	0.36	0.53	0.92	0.2	0.4	0.8
A4	0.08	0.4	0.91	0.4	0.55	0.95	0.06	0.43	0.74	0.27	0.48	0.76	0.18	0.69	0.89	0.06	0.59	0.77	0.1	0.7	0.9
A5	0.21	0.54	0.96	0.24	0.44	0.75	0.07	0.46	0.79	0.29	0.59	0.87	0.19	0.66	0.85	0.18	0.66	0.75	0.3	0.5	0.8
A6	0.25	0.59	0.94	0.17	0.63	0.93	0.13	0.46	0.77	0.27	0.41	0.8	0.23	0.61	0.8	0.13	0.54	0.79	0.3	0.7	0.9
A7	0.33	0.57	0.75	0.32	0.49	0.85	0.36	0.61	0.86	0.19	0.47	0.9	0.1	0.57	0.95	0.32	0.7	0.74	0.4	0.6	0.9
A8	0.1	0.67	0.7	0.04	0.43	0.73	0.08	0.46	0.72	0.35	0.48	0.91	0.17	0.43	0.82	0.1	0.61	0.8	0.4	0.7	1
A9	0.2	0.45	0.98	0.28	0.55	0.75	0.37	0.61	0.86	0.06	0.65	0.74	0.13	0.52	0.94	0.39	0.7	0.88	0.3	0.5	0.8
A10	0.29	0.55	0.83	0.2	0.44	0.71	0.13	0.45	0.88	0.24	0.51	0.93	0.17	0.52	0.8	0.35	0.52	0.88	0.4	0.4	0.7
A11	0.03	0.61	0.71	0.35	0.57	0.96	0.4	0.45	0.77	0.21	0.55	0.95	0.06	0.48	0.72	0.4	0.57	0.97	0.2	0.6	0.7
A12	0.03	0.56	0.72	0.28	0.58	0.93	0.16	0.42	0.89	0.15	0.53	0.97	0.18	0.47	0.78	0.29	0.55	0.88	0.3	0.7	0.7
A13	0.33	0.65	0.9	0.36	0.7	0.7	0.17	0.7	0.81	0.02	0.7	0.75	0.17	0.43	0.84	0.36	0.46	0.81	0.2	0.5	1
A14	0.06	0.6	0.85	0.35	0.58	0.98	0.25	0.44	0.81	0.27	0.58	0.89	0.29	0.47	0.92	0.4	0.52	0.89	0.3	0.5	0.7
A15	0.39	0.6	0.93	0.22	0.54	0.93	0.07	0.63	0.95	0.15	0.59	0.93	0.03	0.52	0.7	0.37	0.7	0.88	0.2	0.7	0.9
A16	0.19	0.53	0.93	0.1	0.55	0.96	0.15	0.61	0.78	0.01	0.42	0.98	0.09	0.4	0.75	0.05	0.41	0.87	0.3	0.5	0.9
A17	0.04	0.44	0.75	0.23	0.66	0.91	0.22	0.65	0.87	0.27	0.47	0.81	0.06	0.48	0.75	0.23	0.69	0.91	0.4	0.4	0.9
A18	0.16	0.65	0.93	0.24	0.47	0.89	0.14	0.49	0.83	0.05	0.48	0.77	0.06	0.58	0.96	0.27	0.56	0.77	0.4	0.5	0.7
A19	0.03	0.52	0.85	0.04	0.59	0.89	0.17	0.51	0.86	0.14	0.44	0.8	0.38	0.46	0.83	0.39	0.56	0.7	0.2	0.7	0.8
A20	0.17	0.6	0.88	0.3	0.67	0.98	0.3	0.53	0.82	0.05	0.67	0.72	0.16	0.56	0.77	0.28	0.56	0.71	0.2	0.6	0.7
A21	0.12	0.53	0.7	0.31	0.58	0.96	0.05	0.4	0.78	0.38	0.52	0.71	0.03	0.53	0.75	0.36	0.5	0.76	0.1	0.7	0.9
A22	0.4	0.45	0.73	0.24	0.4	0.73	0.13	0.6	0.97	0.14	0.62	0.93	0.02	0.69	0.82	0.05	0.49	0.76	0.1	0.7	0.7
A23	0.15	0.46	0.84	0.35	0.55	0.94	0.38	0.54	0.76	0.22	0.61	0.95	0.39	0.63	0.7	0.27	0.42	0.77	0.2	0.6	0.8
A24	0.14	0.69	0.96	0.09	0.57	0.88	0.31	0.63	0.91	0.03	0.49	0.71	0.28	0.61	0.88	0.12	0.67	0.82	0.2	0.6	1
A25	0.03	0.62	0.77	0.02	0.59	0.8	0.3	0.43	0.89	0.08	0.62	0.9	0.23	0.46	0.92	0.31	0.58	0.92	0.3	0.4	0.8

(continued)

Table 4.26 (continued)

	C1			C2			C3			C4			C5			C6			C7		
A26	0.17	0.56	0.97	0.02	0.55	0.75	0.19	0.46	0.72	0.36	0.58	0.72	0.1	0.51	0.95	0.05	0.7	0.94	0.2	0.7	0.8
A27	0.17	0.7	0.78	0.05	0.46	0.74	0.33	0.45	0.74	0.37	0.64	0.78	0.35	0.52	0.79	0.03	0.54	0.79	0.1	0.6	0.8
A28	0.29	0.6	0.85	0.08	0.41	0.88	0.27	0.67	0.84	0.22	0.7	0.9	0.25	0.68	0.96	0.26	0.47	0.86	0.4	0.7	0.9
A29	0.28	0.6	0.75	0.12	0.56	0.9	0.29	0.44	0.97	0.34	0.53	0.83	0.24	0.5	0.94	0.25	0.58	0.83	0.3	0.6	1
A30	0.06	0.7	0.74	0.2	0.56	0.82	0.22	0.61	0.71	0.23	0.48	0.91	0.18	0.68	0.7	0.02	0.55	0.81	0	0.5	0.7
A31	0.02	0.57	0.95	0.05	0.55	0.94	0.33	0.63	0.73	0.21	0.6	0.78	0.22	0.62	0.75	0.05	0.54	0.83	0.3	0.6	0.8
A32	0.27	0.45	0.8	0.35	0.48	0.76	0.22	0.5	0.85	0.06	0.54	0.8	0.14	0.45	0.79	0.23	0.64	0.9	0.1	0.6	0.8
A33	0.19	0.7	0.74	0.23	0.59	0.82	0.16	0.52	0.75	0.32	0.64	0.89	0.17	0.57	0.71	0.35	0.41	0.76	0	0.4	0.7
A34	0.35	0.59	0.8	0.09	0.69	0.72	0.11	0.4	0.96	0.06	0.4	0.86	0.23	0.48	0.77	0.19	0.69	0.92	0.2	0.4	0.8
A35	0.08	0.53	0.83	0.05	0.44	0.74	0.27	0.68	0.74	0.13	0.69	0.98	0.1	0.44	0.97	0.19	0.5	0.71	0.4	0.5	0.7
A36	0.05	0.58	0.76	0.25	0.57	0.71	0.37	0.64	0.86	0.12	0.64	0.95	0.38	0.65	0.91	0.3	0.55	0.75	0.2	0.7	0.8
A37	0.16	0.58	0.77	0.38	0.62	0.91	0.18	0.47	0.91	0.24	0.67	0.97	0.08	0.51	0.75	0.18	0.45	0.91	0	0.4	0.7
A38	0.12	0.59	0.77	0.03	0.66	0.97	0.1	0.41	0.92	0.22	0.62	0.86	0.01	0.49	0.9	0.15	0.69	0.7	0.3	0.5	0.9
A39	0.33	0.7	0.91	0.4	0.66	0.92	0.27	0.62	0.88	0.02	0.53	0.88	0.26	0.56	0.82	0.34	0.59	0.85	0.3	0.6	0.9
A40	0.14	0.58	0.73	0.21	0.45	0.81	0.17	0.52	0.93	0.21	0.51	0.97	0.12	0.55	0.92	0.27	0.62	0.72	0.4	0.6	1
A41	0.37	0.67	0.93	0.06	0.4	0.97	0.13	0.65	0.92	0.34	0.66	0.8	0.31	0.57	0.91	0.36	0.4	0.78	0.2	0.4	0.7
A42	0.11	0.58	0.7	0.13	0.49	0.79	0.35	0.55	0.88	0.24	0.67	0.97	0.26	0.43	0.84	0.08	0.68	0.71	0.3	0.5	0.9
A43	0.18	0.49	0.74	0.19	0.6	0.7	0.39	0.53	0.71	0.27	0.46	0.88	0.14	0.42	0.74	0.24	0.59	0.88	0.3	0.6	0.9
A44	0.08	0.53	0.72	0.34	0.57	0.94	0.35	0.59	0.8	0.03	0.52	0.89	0.08	0.6	0.82	0.35	0.41	0.93	0.2	0.6	0.8
A45	0.24	0.54	0.9	0.14	0.53	0.71	0.4	0.46	0.88	0.38	0.65	0.84	0.28	0.47	0.7	0.22	0.61	0.76	0.4	0.5	0.7
A46	0.28	0.59	0.7	0.08	0.6	0.79	0.25	0.52	0.74	0.31	0.52	0.98	0.22	0.48	0.97	0.22	0.61	0.97	0.1	0.6	0.9
A47	0.03	0.49	0.85	0.36	0.43	0.98	0.02	0.53	0.75	0.4	0.66	0.81	0.37	0.52	0.7	0.18	0.42	0.71	0.4	0.5	0.9
A48	0.27	0.52	0.93	0.22	0.61	0.98	0.3	0.51	0.94	0.19	0.47	0.92	0.27	0.65	0.98	0.26	0.64	0.9	0.2	0.5	0.8
A49	0.29	0.7	0.85	0.12	0.52	0.83	0.3	0.57	0.75	0.36	0.68	0.86	0.03	0.53	0.86	0.14	0.69	0.85	0.2	0.6	0.9
A50	0.14	0.43	0.87	0.31	0.65	0.72	0.39	0.48	0.96	0.24	0.44	0.96	0.28	0.62	0.88	0.39	0.42	0.76	0.1	0.4	0.9

A51	0.32	0.53	0.72	0.08	0.51	0.71	0.09	0.51	0.72	0.19	0.46	0.96	0.3	0.51	0.86	0.32	0.68	0.92	0.2	0.6	0.8
A52	0.11	0.44	0.78	0.21	0.5	0.75	0.26	0.45	0.71	0.31	0.67	0.78	0.05	0.41	0.98	0.12	0.44	0.94	0.1	0.7	1
A53	0.18	0.56	0.83	0.09	0.68	0.89	0.29	0.5	0.89	0.27	0.6	0.73	0.12	0.58	0.97	0.32	0.48	0.76	0.4	0.7	0.7
A54	0.36	0.56	0.97	0.19	0.68	0.73	0.16	0.59	0.7	0.17	0.48	0.9	0.08	0.45	0.79	0.13	0.65	0.93	0.2	0.7	0.8
A55	0.26	0.69	0.76	0.3	0.62	0.86	0.37	0.64	0.91	0.12	0.67	0.93	0.38	0.52	0.77	0.23	0.57	0.78	0.4	0.4	1
A56	0.28	0.48	0.89	0.08	0.58	0.78	0.33	0.51	0.87	0.16	0.55	0.9	0.07	0.52	0.8	0.28	0.64	0.82	0.4	0.6	0.8
A57	0.28	0.42	0.77	0.06	0.46	0.95	0.24	0.56	0.77	0.34	0.58	0.86	0.06	0.53	0.72	0.18	0.54	0.78	0.1	0.6	0.7
A58	0.09	0.6	0.94	0.03	0.47	0.71	0.1	0.54	0.76	0.14	0.54	0.9	0.25	0.4	0.86	0.28	0.68	0.96	0.2	0.7	0.9
A59	0.14	0.64	0.89	0.18	0.4	0.96	0.33	0.7	0.7	0.36	0.62	0.7	0.32	0.47	0.82	0.3	0.48	0.89	0.2	0.6	0.8
A60	0.01	0.58	0.81	0.08	0.42	0.78	0.22	0.42	0.93	0.27	0.53	0.82	0.23	0.41	0.84	0.06	0.43	0.75	0.4	0.7	0.8
A61	0.37	0.4	0.83	0.19	0.43	0.98	0.4	0.42	0.97	0.05	0.65	0.79	0.26	0.46	0.94	0.06	0.58	0.72	0.4	0.5	0.8
A62	0.17	0.54	0.92	0.14	0.49	0.71	0.01	0.61	0.92	0.1	0.5	0.8	0.39	0.66	0.84	0.33	0.62	0.96	0.1	0.5	0.9
A63	0.13	0.64	0.83	0.12	0.41	0.84	0.22	0.67	0.96	0.22	0.57	0.81	0.12	0.63	0.76	0.2	0.6	0.95	0.3	0.6	0.9
A64	0.02	0.45	0.9	0.31	0.59	0.72	0.26	0.44	0.76	0.16	0.55	0.89	0.39	0.59	0.87	0.22	0.48	0.98	0.2	0.6	0.8
A65	0.19	0.44	0.79	0.04	0.64	0.96	0.08	0.41	0.73	0.39	0.62	0.81	0.07	0.42	0.77	0.02	0.5	0.98	0.2	0.4	0.7
A66	0.25	0.45	0.91	0.22	0.43	0.93	0.25	0.69	0.8	0.34	0.44	0.71	0.35	0.68	0.9	0.15	0.49	0.73	0.3	0.6	0.8
A67	0.1	0.68	0.77	0.14	0.49	0.91	0.17	0.7	0.97	0.04	0.45	0.79	0.3	0.47	0.86	0.37	0.44	0.79	0.2	0.6	0.9
A68	0.31	0.45	0.78	0.01	0.41	0.89	0.28	0.7	0.92	0.13	0.4	0.85	0.33	0.52	0.98	0.36	0.55	0.87	0.1	0.5	0.7
A69	0.04	0.57	0.89	0.25	0.56	0.91	0.14	0.6	0.77	0.04	0.44	0.88	0.04	0.49	0.84	0.24	0.61	0.7	0.2	0.6	0.7
A70	0.22	0.53	0.88	0.29	0.64	0.78	0.12	0.61	0.85	0.35	0.7	0.86	0.03	0.62	0.86	0.22	0.4	0.93	0.3	0.4	0.7
A71	0.26	0.61	0.88	0.28	0.57	0.81	0.17	0.58	0.91	0.4	0.57	0.84	0.22	0.65	0.94	0.14	0.66	0.77	0.4	0.7	0.7
A72	0.38	0.46	0.9	0.03	0.64	0.79	0.24	0.57	0.86	0.14	0.53	0.84	0.32	0.49	0.83	0.24	0.69	0.71	0.2	0.7	0.8
A73	0.1	0.43	0.87	0.25	0.62	0.73	0.21	0.42	0.9	0.03	0.67	0.71	0.31	0.43	0.73	0.03	0.4	0.89	0.3	0.7	0.9
A74	0.19	0.54	0.89	0.06	0.57	0.84	0.4	0.5	0.98	0.18	0.65	0.81	0.11	0.56	0.98	0.24	0.43	0.93	0	0.5	1
A75	0.31	0.5	0.89	0.36	0.64	0.91	0.14	0.67	0.83	0.25	0.65	0.95	0.29	0.49	0.78	0.25	0.42	0.72	0.3	0.5	0.9
A76	0.17	0.66	0.94	0.03	0.42	0.72	0.17	0.46	0.73	0.38	0.45	0.77	0.35	0.68	0.88	0.06	0.64	0.72	0.1	0.6	1
A77	0.11	0.59	0.86	0.32	0.69	0.89	0.13	0.62	0.92	0.36	0.58	0.84	0.11	0.42	0.94	0.1	0.47	0.73	0.3	0.5	0.9

(continued)

Table 4.26 (continued)

	C1			C2			C3			C4			C5			C6			C7		
A78	0.22	0.66	0.77	0.06	0.62	0.73	0.28	0.4	0.94	0.25	0.65	0.85	0.24	0.69	0.71	0.35	0.61	0.91	0.2	0.7	0.7
A79	0.13	0.43	0.97	0.05	0.59	0.79	0.37	0.63	0.71	0.09	0.54	0.82	0.24	0.48	0.94	0.27	0.56	0.79	0.1	0.7	0.7
A80	0.26	0.54	0.88	0.27	0.63	0.86	0.16	0.61	0.91	0.39	0.59	0.9	0.08	0.53	0.81	0.27	0.43	0.74	0.2	0.6	0.9
A81	0.22	0.6	0.85	0.3	0.47	0.91	0.09	0.48	0.89	0.29	0.5	0.84	0.34	0.6	0.76	0.01	0.69	0.98	0.1	0.5	0.8
A82	0.29	0.56	0.98	0.39	0.66	0.72	0.2	0.59	0.76	0.23	0.44	0.86	0.13	0.44	0.89	0.05	0.54	0.89	0.2	0.6	0.7
A83	0.09	0.43	0.73	0.15	0.51	0.89	0.08	0.65	0.92	0.28	0.53	0.94	0.23	0.45	0.74	0.12	0.63	0.86	0.4	0.6	0.7
A84	0.03	0.52	0.83	0.24	0.64	0.8	0.38	0.43	0.75	0.3	0.51	0.83	0.2	0.68	0.86	0.18	0.48	0.91	0.2	0.7	0.9
A85	0.15	0.63	0.88	0.09	0.42	0.92	0.04	0.55	0.75	0.16	0.64	0.91	0.02	0.41	0.93	0.36	0.62	0.81	0.3	0.5	0.8
A86	0.31	0.68	0.98	0.09	0.52	0.86	0.36	0.43	0.71	0.18	0.61	0.97	0.19	0.51	0.92	0.38	0.47	0.85	0.2	0.5	0.8
A87	0.08	0.44	0.9	0.1	0.59	0.84	0.23	0.63	0.79	0.32	0.61	0.73	0.15	0.56	0.9	0.39	0.48	0.77	0.4	0.7	0.9
A88	0.04	0.56	0.85	0.07	0.64	0.72	0.08	0.5	0.76	0.16	0.58	0.83	0.35	0.5	0.88	0.38	0.42	0.78	0.4	0.5	0.9
A89	0.35	0.55	0.81	0.12	0.47	0.85	0.21	0.68	0.88	0.03	0.47	0.94	0.39	0.42	0.87	0.24	0.46	0.88	0.3	0.7	0.9
A90	0.27	0.62	0.85	0.04	0.52	0.73	0.05	0.52	0.71	0.01	0.66	0.72	0.17	0.49	0.77	0.32	0.55	0.88	0.2	0.6	0.9
A91	0.14	0.44	0.86	0.05	0.64	0.78	0.21	0.53	0.98	0.27	0.55	0.76	0.31	0.7	0.75	0.32	0.47	0.87	0.3	0.4	1
A92	0.11	0.41	0.91	0.25	0.69	0.82	0.33	0.55	0.95	0.23	0.43	0.89	0.32	0.46	0.98	0.05	0.55	0.94	0	0.6	0.7
A93	0.1	0.53	0.89	0.28	0.63	0.82	0.06	0.52	0.96	0.24	0.41	0.71	0.33	0.53	0.91	0.37	0.55	0.78	0.2	0.7	0.8
A94	0.15	0.62	0.81	0.27	0.43	0.97	0.37	0.62	0.87	0.07	0.4	0.82	0.2	0.65	0.8	0.27	0.68	0.84	0.4	0.6	0.8
A95	0.28	0.61	0.82	0.08	0.48	0.93	0.14	0.69	0.73	0.34	0.59	0.85	0.03	0.58	0.96	0.39	0.46	0.73	0.1	0.5	0.9
A96	0.01	0.5	0.82	0.2	0.63	0.81	0.3	0.6	0.94	0.35	0.43	0.96	0.08	0.53	0.91	0.12	0.52	0.82	0	0.6	0.9
A97	0.11	0.46	0.93	0.11	0.41	0.89	0.16	0.63	0.94	0.05	0.56	0.74	0.02	0.69	0.92	0.13	0.61	0.72	0.3	0.7	0.8
A98	0.18	0.67	0.81	0.18	0.54	0.87	0.13	0.57	0.98	0.23	0.4	0.92	0.23	0.45	0.84	0.17	0.49	0.78	0.1	0.7	0.9
A99	0.31	0.52	0.93	0.03	0.49	0.92	0.22	0.5	0.87	0.34	0.68	0.98	0.21	0.7	0.94	0.05	0.58	0.78	0.2	0.4	0.9
A100	0.31	0.51	0.76	0.28	0.43	0.73	0.15	0.63	0.82	0.21	0.48	0.72	0.39	0.61	0.81	0.07	0.4	0.78	0.1	0.6	0.8

Table 4.27 Best and worst fuzzy values for each criterion

Criteria	C1	C2	C3	C4	C5	C6	C7
Best value	(0, 0.6, 1)	(0, 0.5, 1)	(0.12, 0.57, 1)	(0.36, 0.64, 1)	(0.23, 0.64, 1)	(0.06, 0.78, 1)	(−0.06, 0.46, 1)
Worst value	(−0.5, 0, 0.5)	(−0.49, 0, 0.49)	(−0.42, 0, 0.42)	(−0.38, 0, 0.38)	(−0.35, 0, 0.35)	(−0.47, 0, 0.47)	(−0.43, 0, 0.43)

the names of some techniques under the superiority approach are derived from Greek mythology (Argus, Orestes, Prometheus, and Electra).

Generally, superiority techniques utilize the concept of dominance and preference relations between two alternatives. The dominance relation ($A_1 \rightarrow A_8$) indicates that, mathematically, (A_1) does not completely dominate (A_8); however, considering the dominance of (A_1) over (A_8) carries less risk compared to the dominance of (A_8) over (A_1). In other words, (A_1) is not absolutely superior to (A_8), but the dominance of (A_1) over (A_8) is closer to reality. Typically, symbols such as (R, J, Q, P, S) with the definitions provided in the table below are used to express dominance relations.

A_l S A_h	It is desirable that A_l is at least as good as A_h
A_l I A_h	A_l and A_h are indifferent to each other
A_l Q A_h	A_l has weak dominance over A_h
A_l P A_h	A_l has strong dominance over A_h
A_l R A_h	A_l and A_h are incomparable

In superiority techniques, pairwise comparisons between alternatives are also used; however, they differ in approach compared to pairwise comparison techniques such as AHP and similar methods. In the techniques discussed in Chap. 2, pairwise comparisons form the inputs of a problem, but in superiority techniques pairwise comparisons between alternatives are part of the computational steps for evaluating the alternatives. The pairwise comparisons used in superiority techniques assume the relations A_1SA_h and A_hSA_1, and then it is examined which of the relations A_1SA_h or A_hSA_1 is supported by the available evidence (including problem data such as the decision matrix, weight vector, etc.). This process is performed for all alternatives, and finally all proven assumptions corresponding to the relations R, I, P, and Q are aggregated in the form of a final relation matrix, and decision-making is carried out (Safari et al., 2024). The main idea of most superiority techniques is similar, and their differences lie in the way pairwise comparisons are scored and the final matrix is aggregated. Superiority techniques have three main differences compared to other multi-criteria decision-making techniques:

- Comparing two alternatives at each stage instead of comparing all alternatives simultaneously
- Using a pseudo-criterion instead of a true criterion
- Providing pairwise relations between alternatives instead of ranking them

Table 4.28 Ranking of suppliers based on the total distances from the ideal solution ($\widetilde{S}_i$), regret values ($\widetilde{R}_i$), and compromise measure ($\widetilde{Q}_i$)

Alternatives	$\widetilde{R}_i$	$\widetilde{S}_i$	$\widetilde{Q}_i$
A1	89	58	15
A2	99	60	8
A3	56	91	15
A4	36	3	94
A5	16	5	98
A6	55	47	49
A7	9	76	66
A8	27	40	80
A9	29	88	35
A10	72	43	38
A11	94	49	17
A12	39	27	81
A13	65	84	13
A14	52	34	65
A15	54	80	25
A16	15	61	76
A17	33	62	55
A18	83	98	2
A19	77	79	10
A20	53	96	13
A21	81	45	31
A22	45	19	82
A23	88	46	25
A24	12	36	88
A25	80	51	26
A26	23	75	53
A27	63	95	9
A28	75	64	19
A29	42	39	70
A30	85	89	4
A31	78	2	73
A32	26	100	31
A33	61	66	29
A34	11	28	94
A35	69	71	18
A36	14	54	78
A37	96	41	21
A38	98	6	46
A39	79	65	16
A40	3	44	90
A41	40	63	47

(continued)

Table 4.28 (continued)

Alternatives	$\widetilde{R}_i$	$\widetilde{S}_i$	$\widetilde{Q}_i$
A42	31	67	53
A43	86	22	44
A44	71	23	57
A45	37	16	87
A46	62	68	27
A47	92	92	1
A48	93	20	41
A49	84	18	49
A50	34	29	83
A51	4	87	62
A52	7	26	95
A53	100	53	11
A54	46	1	90
A55	48	25	77
A56	1	15	99
A57	70	93	6
A58	57	21	74
A59	58	9	80
A60	97	38	22
A61	35	81	37
A62	32	77	42
A63	41	50	62
A64	91	30	34
A65	21	69	63
A66	18	59	75
A67	74	97	5
A68	44	12	86
A69	82	52	25
A70	30	13	91
A71	64	17	70
A72	60	78	20
A73	68	48	37
A74	66	31	54
A75	49	11	85
A76	28	85	41
A77	2	90	59
A78	10	73	67
A79	67	24	62
A80	13	10	97
A81	73	7	73
A82	51	70	34
A83	19	82	51

(continued)

Table 4.28 (continued)

Alternatives	$\widetilde{R}_i$	$\widetilde{S}_i$	$\widetilde{Q}_i$
A84	47	42	64
A85	24	83	45
A86	38	56	57
A87	50	74	32
A88	6	55	84
A89	8	4	100
A90	90	72	7
A91	22	86	44
A92	59	33	59
A93	25	57	68
A94	87	14	51
A95	5	35	92
A96	20	94	39
A97	43	37	73
A98	17	8	96
A99	76	99	3
A100	95	32	29

1. **Pairwise Comparison of Alternatives**

The most significant difference between superiority techniques and other techniques is that, in most techniques, alternatives are compared simultaneously. However, superiority techniques compare alternatives pairwise. For example, in the techniques discussed before this chapter, a numerical value is obtained for each alternative in each criterion at each step. In contrast, superiority techniques compare alternatives pairwise from the very beginning or after normalizing and weighting the decision matrix and perform calculations for each pair of alternatives at each stage. In the tables below, you can observe how alternatives are evaluated in non-superiority techniques (Table 4.29) and superiority techniques (Table 4.30) for the decision matrix.

It is evident that the computational volume in superiority techniques is significantly higher than in other techniques, and this difference becomes more pronounced when the number of criteria and alternatives increases. Additionally, the computer implementation of superiority techniques is generally more complex than that of other techniques. Another important point is that, in different steps, the value assigned to the pair of alternatives (A_l, A_h) is not necessarily equal to the value assigned to the pair (A_l, A_h), and they can differ from each other.

2. **Pseudo-Criterion**

The indices used in non-superiority techniques are mostly of the true criterion type. In a true criterion, any difference leads to superiority; in other words, if we

Table 4.29 The method of evaluating options in non-superiority techniques for the decision matrix

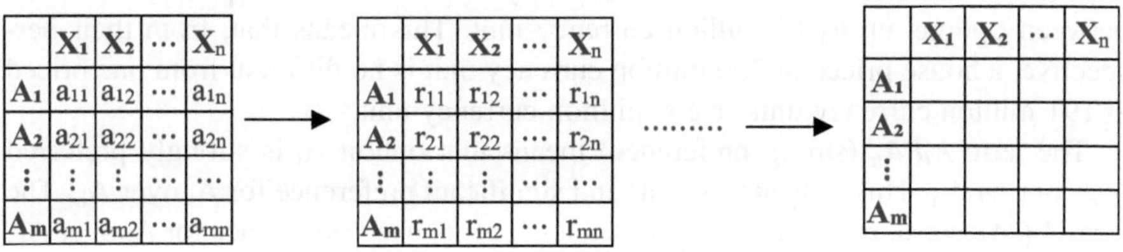

Table 4.30 The method of evaluating options in superiority techniques for the decision matrix

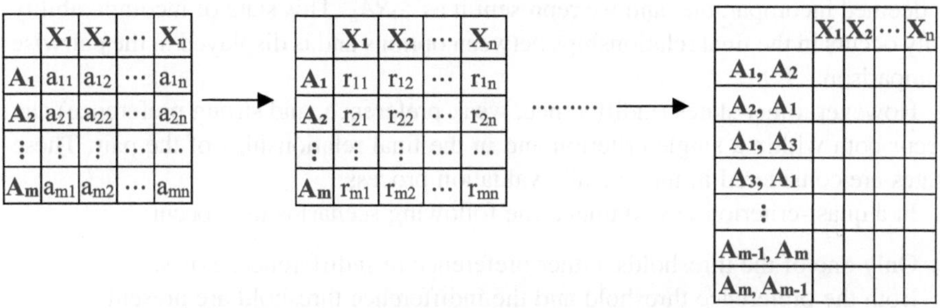

consider a true criterion with a positive nature, 2 is better than 1, 1 which is better than 0.05, which is better than 0.001, which is better than 0.0001.

The concept of a pseudo-criterion stands in contrast to that of a true criterion. In a pseudo-criterion, some differences are considered negligible and can be ignored because they are insignificant from the decision-maker's perspective. For example, in a car purchase decision, the decision-maker might state that they are indifferent to speed differences of up to 25 km/h. This means that, from their perspective, a car with a maximum speed of 180 km/h is indifferent to cars with maximum speeds of 165 km/h or 155 km/h. The number 25 km/h is called the indifference threshold for the speed criterion.

The decision-maker also states that in the difference of vehicle speeds, a range from 20 km/h to 40 km/h is significant for them; however, the degree of importance of this difference is weak preference. Ultimately, a speed difference exceeding 40 km/h becomes a strict preference for the decision-maker (DM). The value of 40 km/h is referred to as the preference threshold with a strong preference (superiority) threshold. Typically, the indifference threshold for the index X_j is denoted by q_j, and the preference threshold is denoted by p_j.

For negative indices, this axis will be reversed; for example, the price of option A_l must be lower than that of option A_h to be preferred. For instance, in a house-buying problem, the decision-maker states that they are indifferent to price differences between options up to 15 million currency unit. This means that, from their perspective, a house priced at 200 million currency unit is no different from one priced at 191 million currency unit or 209 million currency unit.

The term A_lPA_h (strong preference) means that option A_l is strongly preferred over option A_h. This indicates a clear and significant preference for A_l over A_h. The term A_lQA_h (weak preference) suggests a less pronounced preference for A_l over A_h. The term A_lIA_h (indifference) means that there is no preference between the two options, indicating that they are considered equal in value.

A superiority technique may include some or all of these states. Another state exists where, based on the data, it is not possible to declare a superiority or indifference relationship between options A_l and A_h. In this case, the comparison is deemed incomparable, and we represent it as A_lRA_h. This state of incomparability only occurs in the final relationships between options and is displayed in the pairwise comparison.

However, other states (indifference, weak preference, and strong preference) can occur both within a single criterion and in the final relationships of the pair. These states are considered in the overall evaluation process.

In a quasi-criterion environment, the following scenarios may occur:

1. Only one of the thresholds, either preference or indifference, exists.
2. Both the preference threshold and the indifference threshold are present.
3. The utility changes uniformly, monotonically, or in any other manner within the intervals between the thresholds.

These scenarios help in understanding how different thresholds and utility changes can influence decision-making processes in various contexts.

An important point regarding the detection thresholds is that they can be presented in both additive and multiplicative forms. The additive form means that, for example, if in the price index A_l is 10 million or 30 million less than A_h, then A_l has an advantage over A_h. The multiplicative form means that, for example, if in the price index A_l is less than 0.7 times the price of A_h, then A_l has an advantage over A_h, or if the speed of vehicle A_l is more than 1.3 times that of vehicle A_h, then A_l has an advantage over A_h. In this book, all detection thresholds are presented in the additive form.

3. **Pairwise Relationships**

Given that superiority techniques compare options in pairs rather than examining them individually, their output also reflects the pairwise relationships between the options. For example, in other techniques, the output includes a ranking of the options. However, the output of superiority techniques indicates that, based on the input of the problem, option A_1 has an advantage over option A_2, and A_2 has an advantage over option A_3, and they typically do not rank the options. The tables below illustrate how the output is extracted in superiority and non-superiority techniques (Tables 4.31 and 4.32).

Additionally, the output of superiority techniques can be represented in a pairwise comparison matrix, where each entry indicates the superiority of the row option over the column option.

	A_1	A_2	A_3
A_1	–	✓	
A_2		–	✓
A_3			–

This matrix indicates that:

A_1 has superiority over A_2 (the check mark in row A_1 and column A_2).
A_2 has superiority over A_3 (the check mark in row A_2 and column A_3).
There is no superiority for A_3 over A_1 or A_2 (the dashes in row A_3).

Additionally, the final relationship matrix can be represented as a superiority diagram. This diagram is a directed graph where the problem's options are represented by nodes, and the superiority relationships are indicated by arrows pointing from the superior option to the inferior option.

Table 4.31 The method of extracting output in non-superiority techniques

	X_1	X_2	X_3
A_1	a_{11}	a_{12}	a_{13}
A_2	a_{21}	a_{22}	a_{23}
A_3	A_{31}	a_{32}	a_{33}

→

	X_1	X_2	X_3
A_1	r_{11}	r_{12}	r_{13}
A_2	r_{21}	r_{22}	r_{23}
A_3	R_{31}	r_{32}	r_{33}

→

	X_1	X_2	X_n
A_1			
A_2			
A_m			

→

	Score	Rank
A_1	0.2	3
A_2	0,7	1
A_m	0,6	2

Table 4.32 The method of extracting output in superiority techniques

	X_1	X_2	X_3
A_1	a_{11}	a_{12}	a_{13}
A_2	a_{21}	a_{22}	a_{23}
A_3	A_{31}	a_{32}	a_{33}

→

	X_1	X_2	X_3
A_1	r_{11}	r_{12}	r_{13}
A_2	r_{21}	r_{22}	r_{23}
A_3	R_{31}	r_{32}	r_{33}

→

	X_1	X_2	X_n
A_1, A_2			
A_2, A_1			
A_1, A_3			
A_3, A_1			
A_2, A_3			
A_3, A_2			

→

	Superiority Confirmation
A_1, A_2	✓
A_2, A_1	
A_1, A_3	
A_3, A_1	
A_2, A_3	✓
A_3, A_2	

4.10.1 The ELECTREE Method

The ELECTRE (elimination and choice translating reality) method, also known as the outranking method, is one of the multi-criteria decision-making (MCDM) techniques. Its name is derived from the French phrase "Élimination et Choix Traduisant la Réalité," meaning "elimination and choice expressing reality." Various articles have provided examples of its applications, including project ranking, facility layout planning, factory location selection, strategic planning, and more.

The questionnaire for this method is similar to that of the TOPSIS method, consisting of a criteria-alternative matrix. Responses are collected using linguistic scales or real numbers.

This method was first introduced by Mr. Benayoun in 1966 and later developed by researchers Roy and Van Delft. In this method, all alternatives are evaluated using non-ranking comparisons, and ineffective alternatives are eliminated. These steps are based on concordance and discordance sets, which is why this method is also known as concordance analysis. The ELECTRE method has various versions, including ELECTRE I, II, III, and IV. The differences between these versions lie in the mathematical operations and the types of problems they can solve.

The core of this method is based on non-ranking relationships; therefore, the results obtained are presented as a set of rankings. Decision-making problems are modeled using this method by forming a decision matrix, where the rows represent competing alternatives and the columns represent decision criteria.

In selection problems, the decision-maker (DM) seeks to identify one or more reference (selected) options, and the status of other options is irrelevant.

In ranking problems, options are divided into nominal groups that are not predetermined, and their superiority relative to each other is determined. In classification problems, options are divided into several predefined ordinal groups. The applications of various techniques in the ELECTRE family are presented in the table below (Table 4.33).

Table 4.33 Applications of various techniques in the ELECTRE family

	Selection of one or more superior options	Ranking of options	Classification of options
ELECTRE I	✓		
ELECTRE Iv	✓		
ELECTRE II		✓	
ELECTRE III		✓	
ELECTRE TRI			✓

4.10.1.1 The Computational Steps of ELECTRE I Are as Follows:

Step 1. Normalize the decision matrix using the Euclidean method.
Step 2. Obtain the weighted normalized matrix by multiplying the weights of the criteria by the corresponding column:

$$t_{ij} = r_{ij} \times w_j.$$

Step 3. Form the concordance set and the discordance set. Consider all pairs of options. For each pair of options (A_l, A_h), two sets are defined as follows. Note that the number of each of the concordance and discordance sets is equal to m (m−1):

(a) Concordance set: This includes the key criteria for which the relation A_1SA_h holds. This relation indicates that option A_l has a preference greater than or equal to option A_h. The concordance set of the option pair (A_l, A_h), denoted by $j^+_{l,h}$, is obtained for positive criteria using relation (I) and for negative criteria using relation (II):

$$j^+_{l,h} = \{x_J | t_{l,j} \geq t_{hj}\} \qquad (I)$$
$$j^+_{l,h} = \{x_J | t_{l,j} \leq t_{hj}\} \qquad (II).$$

(b) Inconsistency set: This includes all indicators where the relation A_1SA_h does not hold. This relation states that option A_l is less preferred than option A_h medication; in other words, the inverse relation of A_1SA_h (i.e., the relation not (A_1SA_h)) holds.

The inconsistency set for the pair of options (A_l, A_h), which we denote by $j^-_{l,h}$, is derived using relation (III) for positive indicators and relation (IV) for negative indicators:

$$j^-_{l,h} = \{x_J | t_{l,j} > t_{hj}\} \qquad (III)$$
$$j^-_{l,h} = \{x_J | t_{l,j} < t_{hj}\} \qquad (IV).$$

When $j^+_{l,h} \cup j^-_{l,h}$, the concordance and discordance sets for each pair of options are mutually exclusive. If the pairs of options (A_l, A_h) and (A_h, A_l) do not have identical values in any of the indicators, it can be said that $j^+_{l,h} \cup j^+_{h,j} = J$ and $j^-_{j,h} \cup j^-_{h,l} = J$.

If this pair of options has identical values in even one indicator, and we denote the set of indicators with identical values as $j^=_{l,h}$, it can be stated that $j^+_{l,h} \cup j^+_{h,j} = J$ and $j^+_{l,h} \cap j^+_{h,j} = j^=_{l,h} = j^=_{h,j}$ and $j^-_{l,h} \cup j^-_{h,j} = J - j^=_{l,h}$ holds.

In other words, if a pair of options has identical values in one indicator, the indicators with identical values will appear in both $j^+_{l,h}$ and $j^+_{h,j}$, but they will not appear in $j^-_{l,h}$ *and* $j^-_{h,l}$.

Step 4. Obtain the concordance matrix and the concordance threshold. The elements of the concordance matrix, denoted by $C_{l,\ h}$, represent the concordance of the pair of options (A_l, A_h) and are equal to the sum of the weights of the indicators in the concordance set of this pair of options $j^+_{l,h}$:

$$C_{l,h} = \sum_{j \in j^+_{l,h}} w_j.$$

If a pair of options does not have identical values in any indicator, the sum of each element of the concordance matrix and its corresponding element with respect to the main diagonal of this matrix equals one; that is, $C_{l,\ h} + C_{h,\ l} = 1$. If they have identical values in even one indicator, we have $C_{l,\ h} + C_{h,\ l} > 1$.

The concordance threshold is an arbitrary value between zero and one, which can be determined either by consulting experts or by considering a critical point from the concordance matrix. The most common critical point used as the threshold is the average of the elements of the concordance matrix, calculated as follows:

$$\overline{C} = \frac{\sum \sum C_{l,h}}{m(m-1)}.$$

Step 5. Obtain the discordance matrix and discordance threshold. The elements of the discordance matrix, denoted by $D_{l,\ h}$, represent the discordance of the pair of options (A_1, A_2) and are derived from the following relation:

$$D_{l,h} = \frac{\max_{j \in J^-_{l,h}} |t_{lj} - t_{hj}|}{\max_{j \in J} |t_{lh} - t_{hj}|}.$$

The discordance threshold is an arbitrary value between 0 and 1, which can be determined either by consulting experts or by considering a critical point from the discordance matrix. The most common critical point used as the threshold is the average of the elements of the discordance matrix, calculated as follows:

$$\overline{D} = \frac{\sum \sum D_{l,h}}{m(m-1)}.$$

Step 6: **Construct the concordance dominance matrix.** If the elements of the concordance matrix are greater than or equal to the concordance threshold, the corresponding entries in the effective concordance matrix will be equal to 1; otherwise, they will be equal to 0. In other words:

$$E_{l,h} = \begin{cases} 1 & C_{l,h} \geq \overline{C} \\ 0 & C_{l,h} < \overline{C} \end{cases}.$$

Step 7. Construct the discordance dominance matrix. If the elements of the discordance matrix are less than or equal to the discordance threshold, the corresponding entries in the effective discordance matrix will be equal to 1; otherwise, they will be equal to 0. In other words:

$$F_{l,h} = \begin{cases} 1 & D_{l,h} \geq \overline{D} \\ 0 & D_{l,h} < \overline{D} \end{cases}.$$

Step 8. Construct the aggregate dominance matrix. Each element of this matrix is obtained by multiplying the corresponding elements of the concordance dominance matrix (E) and the discordance dominance matrix (F):

$$G_{l,h} = E_{l,h} \times F_{l,h}.$$

Step 9. Remove the less favorable options. The options with the lowest number of zeros in their column (with the highest column sum) are considered as the less favorable options. The options with the highest number of zeros in their column (with the lowest column sum) are selected as the preferred options.

Example 4.3

In selecting a mobile phone brand, four different brands within a specific price range have been introduced. The decision-maker aims to evaluate these brands based on five criteria: useful life, aesthetics, weight (heaviness), durability, and hardware specifications. The evaluations of each phone model according to the criteria, along with the weights assigned to each criterion, are presented in the table below. Using the ELECTRE 1 technique, identify the best mobile phone.

DM	Criterion	Useful life	Aesthetic	Heaviness	Durability	Hardware
Alternative		X_1^+	X_2^+	X_3^-	X_4^+	X_5^+
Xperia Z3	A_1	Relatively very high	Relatively high	152	Low	Low
Galaxy S6	A_2	Relatively low	Relatively low	138	Relatively very high	Relatively very low
HTC M9	A_3	Relatively high	Moderate	157	Very low	Moderate
Nexus 6	A_4	Relatively very high	Very low	184	High	Very high
W_j		0.15	0.2	0.15	0.2	0.3

Solution It is first necessary to quantify the qualitative indicators. Using a ten-point scale, the decision matrix is quantified as follows:

DM	Criterion	Useful Life	Aesthetic	Heaviness	Durability	Hardware
Alternative		X_1^+	X_2^+	X_3^-	X_4^+	X_5^+
Xperia Z3	A_1	8	6	152	3	3
Galaxy S6	A_2	4	4	138	8	2
HTC M9	A_3	6	5	157	1	5
Nexus 6	A_4	8	1	184	7	9
W_j		0.15	0.2	0.15	0.2	0.3

Step 1: Normalization of the decision matrix

NDM	Criterion	Useful Life	Aesthetic	Heaviness	Durability	Hardware
Alternative		X_1^+	X_2^+	X_3^-	X_4^+	X_5^+
Xperia Z3	A_1	0.5963	0.6794	0.4791	0.2705	0.2750
Galaxy S6	A_2	0.2981	0.4529	0.4350	0.7213	0.1833
HTC M9	A_3	0.4472	0.5661	0.4949	0.0902	0.4583
Nexus 6	A_4	0.5963	0.1132	0.5800	0.6312	0.825
W_j		0.15	0.2	0.15	0.2	0.3

Step 2: Calculation of the weighted normalized matrix

WNDM	Criterion	Useful life	Aesthetic	Heaviness	Durability	Hardware
Alternative		X_1^+	X_2^+	X_3^-	X_4^+	X_5^+
Xperia Z3	A_1	0.0894	0.1359	0.0719	0.0541	0.0825
Galaxy S6	A_2	0.0447	0.0906	0.0652	0.01443	0.0550
HTC M9	A_3	0.0671	0.0226	0.0870	0.1262	0.2475
Nexus 6	A_4	0.0894	0.0226	0.0870	0.01262	0.2475

Step 3: Formation of concordance and discordance sets. The number of each set should be equal to m(m−1), which is 4(4−1) = 12.

$J_{1,2}^+ = \{X_1, X_2, X_5\}$	$J_{1,2}^- = \{X_3, X_4\}$
$J_{2,1}^+ = \{X_3, X_4\}$	$J_{2,1}^- = \{X_1, X_2, X_5\}$
$J_{1,3}^+ = \{X_1, X_2, X_3, X_4\}$	$J_{1,3}^- = \{X_5\}$
$J_{3,1}^+ = \{X_5\}$	$J_{3,1}^- = \{X_1, X_2, X_3, X_4\}$
$J_{1,4}^+ = \{X_1, X_2, X_3\}$	$J_{1,4}^- = \{X_4, X_5\}$
$J_{4,1}^+ = \{X_1, X_4, X_5\}$	$J_{4,1}^- = \{X_2, X_3\}$
$J_{2,3}^+ = \{X_3, X_4\}$	$J_{2,3}^- = \{X_1, X_2, X_5\}$
$J_{3,2}^+ = \{X_1, X_2, X_5\}$	$J_{3,2}^- = \{X_3, X_4\}$
$J_{2,4}^+ = \{X_2, X_3, X_4\}$	$J_{2,4}^- = \{X_1, X_5\}$
$J_{4,2}^+ = \{X_1, X_5\}$	$J_{4,2}^- = \{X_2, X_3, X_4\}$
$J_{3,3}^+ = \{X_2, X_3\}$	$J_{3,4}^- = \{X_1, X_4, X_5\}$
$J_{4,3}^+ = \{X_1, X_4, X_5\}$	$J_{4,3}^- = \{X_2, X_3\}$

Step 4: Computing the concordance matrix. The concordance matrix is obtained by summing the weights of the relevant compatible sets.

C	A_1	A_2	A_3	A_4
A_1	–	0.65	0.7	0.5
A_2	0.35	–	0.35	0.55
A_3	0.3	0.65	–	0.35
A_4	0.65	0.45	0.65	–

Note: Options A_1 and A_4 in the index X_1 exhibited significant responsibility values; thus, the concordance matrix reflects $C_{1,\,4} + C_{4,\,1} = 0.5 + 0.65 > 1$.

If the desired concordance threshold is selected as the critical mean point, we obtain:

$$\overline{C} = \frac{0.65 + 0.7 + \ldots + 0.45 + 0.65}{4 \times 3} = 0.5125.$$

Step 5: Computing of the discordance matrix

Note that in the computational relationship of the inconsistency degree, the numerator and denominator expressions are nearly identical, with the key distinction that the numerator is calculated for inconsistency-aggregated indices (A_l, A_h), while the denominator is computed across all indices:

D	A_1	A_2	A_3	A_4
A_1	–	1	1	1
A_2	0.5023	–	0.6536	1
A_3	0.6557	1	–	1
A_4	0.6862	0.3529	0.8234	–

$$\overline{D} = \frac{1 + 1 + \ldots + 0.3529 + 0.8234}{4 \times 3} = 0.8062.$$

Step 6: Computation of the concordance dominance matrix.

E	A_1	A_2	A_3	A_4
A_1	–	1	1	0
A_2	0	–	0	1
A_3	0	1	–	0
A_4	1	0	1	–

Step 7: Computation of the discordance dominance matrix.

F	A_1	A_2	A_3	A_4
A_1	–	0	0	0
A_2	1	–	1	1
A_3	1	0	–	0
A_4	1	1	0	–

Step 8: Computation of the aggregate dominance matrix

G		A_1	A_2	A_3	A_4
Xperia Z3	A_1	–	0	0	0
Galaxy S6	A_2	0	–	0	0
HTC M9	A_3	0	0	–	0
Nexus 6	A_4	1	0	0	–
Number of zeros		2	3	3	3

Step 9: Elimination of less favorable alternatives

Option $\mathbf{A_1}$ is identified as the **least attractive alternative** (assigned the value 1 in its column).

Options $\mathbf{A_2}$**,** $\mathbf{A_3}$, and $\mathbf{A_4}$ exhibit the **highest number of zeros** in their respective columns (with equal counts), thus qualifying as the **optimal choices**.

The table below shows zero-count dominance determines attractiveness ranking.

Option	Zero count	Attractiveness
A_1	1	Low
A_2	max(z)	High
A_3	max(z)	High
A_4	max(z)	High

4.10.1.2 The ELECTRE Iv Technique

The ELECTRE Iv technique is an extension of the ELECTRE I method within the ELECTRE family, incorporating a **veto condition**. Although its procedural steps are similar to ELECTRE I, this technique introduces key differences by adding the veto constraint.

The **veto rule** explicitly states that if alternative $\mathbf{A_1}$ outperforms alternative $\mathbf{A_h}$ in terms of the aggregate criteria, but $\mathbf{A_h}$ dominates $\mathbf{A_1}$ in even a single criterion and the performance difference between them exceeds a predefined **veto threshold,** then the superiority of $\mathbf{A_1}$ over $\mathbf{A_h}$ is rejected. In other words, a significantly strong dominance of $\mathbf{A_h}$ in just one criterion can override the relatively weaker dominance of $\mathbf{A_1}$ in other criteria, thereby neutralizing its overall advantage. However, this does not imply that $\mathbf{A_h}$ is superior to $\mathbf{A_1}$.

The key differences between ELECTRE I and ELECTRE Iv are summarized in the table below:

Condition	ELECTRE I	ELECTRE 1v
Overall dominance	✓	✓
Veto threshold	✗	✓
Compensation logic	Full	Limited

4.10.1.3 The ELECTRE II Technique

This technique was proposed by **Roy and Bertier (1972)**. Unlike the two previous methods in the ELECTRE family, which were only capable of identifying the best alternative(s), **ELECTRE II** also has the ability to **rank alternatives**. By introducing **two additional constraints** in the concordance condition, this technique prevents the divergence and selection of multiple alternatives as top-ranked options, thereby eliminating the possibility of indifference between two alternatives.

4.10.1.4 The ELECTRE III Technique

This technique was developed by Roy (1978) to enhance decision-making precision through outranking methods. It should be noted that, generally speaking, techniques that do not treat the outranking of alternatives in a binary (0–1) manner—such as outranking methods employing discrimination thresholds—allow for degrees of outranking between 0 and 1. These methods inherently embody **fuzzy** concepts, **even when fuzzy numbers are not explicitly used in their computations**. ELECTRE III is one such technique.

4.10.1.5 The ELECTRE TRI Technique

The initial concept of the ELECTRE TRI technique was introduced in Yu's doctoral dissertation. The objective of this technique is to classify alternatives into predefined categories. Solving a problem using ELECTRE TRI is similar to solving it with ELECTRE III, with the key difference that alternatives are not compared pairwise; rather, each alternative is compared against one or several boundary alternatives.

These boundary alternatives, denoted as A_y^* where $y = 1, 2, \ldots, z$, are artificial alternatives introduced to the problem that represents the thresholds between categories. When b represents the number of desired categories, z = b−1 boundary alternatives are required. Both the discrimination thresholds and boundary alternatives are determined by domain experts.

The boundary alternatives must maintain a strict ordinal relationship across all criteria. For instance, given two boundary alternatives A_1^* must demonstrate strictly inferior performance to A_2^* on all criteria. Here, inferior performance is defined as lower values for benefit criteria and higher values for cost criteria.

This technique, similar to ELECTRE III, employs pseudo-criteria instead of true criteria. Consequently, normalization is not required.

It is important to note that, in this technique, the indices l and h do not represent pairs of alternatives, as alternatives are only compared against boundary profiles. Thus, the indices l and h are used once for the main alternatives and once for the boundary profiles.

For example, for alternative A_1 and boundary profile A_3^*, the term $C_{l,h}$ refers to two distinct values: $C_{1,\ 3*}$ (comparison of alternative A_1 with boundary A_3^*) and $C_{3\ *\ ,\ 1}$ (comparison of boundary A_3^* with alternative A_1).

4.10.1.6 Case Study

Assessment of Internal and External Consistency of Intuitive Decision-Making Methods in Comparison with Three Multi-criteria Decision-Making Methods (Safari et al., 2024)

Note: The subject of this study is the same as Case Study 4.6.1, which was solved using the fuzzy TOPSIS. The same facility location problem for oil field projects, previously addressed with ELECTRE III, is elaborated in this section. Therefore, based on the steps of the ELECTRE method as previously described, the weighted normalized matrix of the fuzzy TOPSIS method (Table 4.34) is employed, with the procedure being continued from the third step onward.

Step 3: Formation of concordance and discordance sets (Table 4.35)

Step 4: Computing the concordance matrix (Table 4.36)

$$\overline{C} = 0.980$$

Step 5: Computing of the discordance matrix (Table 4.37)

$$\overline{D} = 0.644$$

Step 6: Computation of the concordance dominance matrix (Table 4.38)

Step 7: Computation of the discordance dominance matrix (Table 4.39)

Step 8: Computation of the aggregate dominance matrix and remove the less favorable options (Tables 4.40 and 4.41).

4.10.2 The PROMETHEE Technique

The PROMETHEE technique and the GAIA module are among the most widely used methods in the field of multi-criteria decision-making (MCDM). The PROMETHEE I methodology, which provides a partial ranking, and PROMETHEE II, which offers a complete ranking, were first developed in 1982 by Professor Brans and quickly gained widespread application in its early years. A few years later, Professor Brans, in collaboration with Mareschal, introduced two newer versions of

Table 4.34 Weighted normalizing matrix and positive ideal solution and negative ideal solution

WN	C			EP			DS			FS			IA			HR			DB			PC		
P1	0.4	0.8	1.4	0.0	0.1	0.2	0.0	0.1	0.2	0.0	0.2	0.3	0.4	0.5	0.7	0.2	0.3	0.4	0.4	0.6	0.8	0.3	0.4	0.5
P2	0.1	0.4	0.8	0.0	0.0	0.1	0.0	0.0	0.1	0.0	0.1	0.2	0.6	0.8	1.0	0.1	0.2	0.3	0.3	0.4	0.6	0.1	0.2	0.3
P3	0.3	0.8	1.4	0.2	0.3	0.5	0.3	0.4	0.6	0.1	0.2	0.3	0.4	0.5	0.7	0.3	0.4	0.6	0.2	0.3	0.5	0.6	0.7	0.9
P4	0.4	0.9	1.5	0.1	0.2	0.3	0.2	0.3	0.4	0.2	0.3	0.5	0.6	0.8	1.0	0.1	0.2	0.3	0.3	0.4	0.6	0.0	0.0	0.2
P5	0.2	0.5	0.9	0.1	0.3	0.4	0.2	0.3	0.4	0.1	0.3	0.4	0.2	0.4	0.6	0.2	0.4	0.5	0.0	0.1	0.2	0.6	0.7	0.9
P6	0.3	0.7	1.2	0.1	0.2	0.3	0.3	0.4	0.6	0.0	0.1	0.2	0.4	0.5	0.7	0.1	0.1	0.2	0.3	0.4	0.6	0.0	0.0	0.2
I^+	0.4	0.9	1.5	0.2	0.3	0.5	0.3	0.4	0.6	0.2	0.3	0.5	0.6	0.8	1.0	0.3	0.4	0.6	0.4	0.6	0.8	0.6	0.7	0.9
I^-	0.1	0.4	0.8	0.0	0.0	0.1	0.0	0.0	0.1	0.0	0.1	0.2	0.2	0.4	0.6	0.1	0.1	0.2	0.0	0.1	0.2	0.0	0.0	0.2

Table 4.35 Concordance and discordance sets

	Concordance sets	Discordance sets
A1 and A2	C1, C3, C5, C6, C7	C2, C4, C8
A1 and A3	C1	C2, C3, C7, C8
A1 and A4	C1, C4, C5, C6, C7	C2, C3, C8
A1 and A5	C1, C5, C6, C7	C2, C3, C4, C8
A1 and A6	C1, C4, C5, C6, C7	C2, C3, C8
A2 and A1	C2, C8	C1, C3, C4, C5, C6, C7
A2 and A3	C1	C2, C3, C4, C5, C6, C7, C8
A2 and A4	C1, C2, C5, C6	C3, C7, C8
A2 and A5	C1, C5	C3, C8
A2 and A6	C1, C2, C5	C3, C6, C8
A3 and A1	C2, C3, C7, C8	C1
A3 and A2	C2, C3, C4, C5, C6, C7, C8	C1
A3 and A4	C2, C3, C4, C5, C6, C7, C8	C1
A3 and A5	C2, C3, C4, C5, C6, C7	C1
A3 and A6	C2, C4, C5, C6, C7	C8
A4 and A1	C2, C3, C8	C1, C4, C5, C6, C7
A4 and A2	C3, C7, C8	C1, C2, C5, C6
A4 and A3	C2, C3, C4, C5, C6, C7, C8	C1
A4 and A5	C3, C7	C1, C2, C6, C8
A4 and A6	C2, C7	C3, C6, C8
A5 and A1	C2, C3, C4, C8	C1, C5, C6, C7
A5 and A2	C3, C8	C1, C5
A5 and A3	C1	C2, C3, C4, C5, C6, C7
A5 and A4	C2, C3, C6, C8	C3, C7
A5 and A6	C1, C2	C3, C6, C8
A6 and A1	C2, C3, C8	C1, C4, C5, C6, C7
A6 and A2	C3, C6, C8	C1, C2, C5
A6 and A3	C8	C2, C4, C5, C6, C7
A6 and A4	C3, C6, C8	C2, C7
A6 and A5	C3, C6, C8	C1, C2

Table 4.36 Concordance matrix

C	A1	A2	A3	A4	A5	A6
A1	_	1.19	0.01	1.43	0.94	1.43
A2	0.65	_	0.01	0.85	0.65	0.81
A3	1.15	2.32	_	2.32	1.83	1.58
A4	0.94	0.99	2.32	_	0.5	0.41
A5	1.39	0.74	0.01	0.98	_	0.17
A6	0.94	0.78	0.49	0.78	0.78	_

Table 4.37 Discordance matrix

D	A1	A2	A3	A4	A5	A6
A1	_	1.000	1.000	0.889	1.000	1.000
A2	1.000	_	1.000	1.000	1.000	1.000
A3	0.200	0.179	_	0	0	0.198
A4	0.198	0.343	0.000	_	1.000	1.000
A5	0.675	0.281	1.000	0.422	_	1.000
A6	0.596	0.268	1.000	0.293	0.769	_

Table 4.38 Concordance dominance matrix

E	A1	A2	A3	A4	A5	A6
A1	_	1	0	1	0	1
A2	0	_	0	0	0	0
A3	1	1	_	1	1	1
A4	0	1	1	_	0	0
A5	1	0	0	1	_	0
A6	0	0	0	0	0	_

Table 4.39 Discordance dominance matrix

F	A1	A2	A3	A4	A5	A6
A1	_	1	1	1	1	1
A2	1	_	1	1	1	1
A3	0	0	_	0	0	0
A4	0	0	0	_	1	1
A5	1	0	1	0	_	1
A6	0	0	1	0	1	_

Table 4.40 Aggregate dominance matrix and number of zeros

G = F*E	A1	A2	A3	A4	A5	A6
A1	_	1	0	1	0	1
A2	0	_	0	0	0	0
A3	0	0	_	0	0	0
A4	0	0	0	_	0	0
A5	1	0	0	0	_	0
A6	0	0	0	0	0	_
Number of zeros	4	4	5	4	5	4

Table 4.41 Ranking of alternatives

H	Win	Loss	Different	Rank
A1	3	2	1	1
A2	0	5	−5	3
A3	0	5	−5	3
A4	0	5	−5	3
A5	1	4	−3	2
A6	0	5	−5	3

PROMETHEE: PROMETHEE III (interval-based ranking) and PROMETHEE IV (continuous cases). Additionally, in 1998, the same researchers developed the GAIA visual module, which provides an insightful graphical representation to support PROMETHEE analysis.

The PROMETHEE method has been well received by academics and researchers, with numerous papers utilizing this technique. It is used to evaluate and prioritize discrete alternatives and select the best option based on multiple criteria (with varying measurement scales). When applying the PROMETHEE technique, there is a limitation in compensating for the weakness of one criterion with the strength of another; thus, an ideal alternative must meet at least the minimum requirements across all criteria.

Moreover, the PROMETHEE method can easily handle criteria with different measurement scales (without requiring scale normalization) and defines six distinct preference functions based on the nature of the data and criteria scales. This makes it particularly advantageous in multi-criteria decision-making problems, where criteria often have heterogeneous measurement scales, serving as a strong point for decision-makers.

In a study conducted to select the best statistical distribution using the PROMETHEE and GAIA methods, the results demonstrated that PROMETHEE is a reliable approach for determining the optimal statistical distribution, and its use was strongly recommended. Another study focused on implementing PROMETHEE in multi-criteria decision-making (MCDM). The research confirmed that PROMETHEE serves as an effective tool for evaluating logistics competitiveness among Balkan Peninsula countries (Safari et al., 2014).

In a separate study combining the AHP (analytic hierarchy process) and PROMETHEE approaches to select the most suitable policy for promoting a clean vehicle fleet, researchers found that integrating these two methods enables a precise evaluation of scenarios, identifying and ranking their strengths and weaknesses. Selecting a sustainable concept is a critical challenge for modern manufacturing organizations, as the decision-making process involves multiple criteria. A study employing PROMETHEE successfully identified the best sustainable concept based on social, economic, and environmental criteria (Safari et al., 2014).

4.10.2.1 Computational Steps of the PROMETHEE I Technique

Step 1: Consider all pairs of alternatives and calculate the difference between their values for each criterion separately. The difference between alternative pair (A_l, A_h) for criterion X_i is denoted by $d_{l,\,h}(j)$. Note that the number of alternative pairs (comparisons) equals m(m−1).

Step 2: Calculate the preference value (preference degree) for the differences obtained in Step 1, according to the preference function defined for each criterion. The types of preference functions can be summarized as follows (Table 4.42):

For indicators with a positive nature, the value d is substituted into the preference function; however, for negative indicators, the value −d is substituted into the

Table 4.42 The types of preference functions

Type	Description	Condition	Parameter	Preference function
I (usual criterion)	No indifference area. Absence of difference in interval (d ≤0), the existence of a complete priority of an alternative in interval (d > 0)	Impacts and the issues related to the ecology	–	$P(d)=\begin{cases}0 & d\le 0\\ 1 & d>0\end{cases}$
II (quasi-criterion)	Includes indifference threshold q Lack of differences in the interval (d ≤ q), the existence of a complete priority of an alternative in interval (d > q)	Attributes related to the discrete sources	q	$P(d)=\begin{cases}0 & d\le q\\ 1 & d>q\end{cases}$
III (V-shaped criterion or linear preference)	Includes preference threshold p. Absence of difference in the interval (d ≤ p), the existence of a complete priority of an alternative in interval (d > p)	Operational attributes, purchase costs	p	$P(d)=\begin{cases}0 & \text{if } d\le 0\\ \dfrac{d}{p} & \text{if } 0<d\le p\\ 1 & \text{if } d>p\end{cases}$
IV (level criterion)	Includes both indifference (q) and preference (p) thresholds. Lack of difference in interval (d ≤ q), change in priority value of alternative linearly in the interval (q < d ≤ p), the existence of a complete priority of an alternative in the interval (d > p)	Long-term benefit, maintenance cost, lifetime cost	P, q	$P(d)=\begin{cases}0 & \text{if } d\le q\\ \dfrac{1}{2} & \text{if } q<d\le p\\ 1 & \text{if } d>p\end{cases}$
V (linear criterion with indifference area)	Similar to type III but with indifference threshold q. Absence of difference in the interval (d ≤ q), change in the priority value of alternative	Exploration cost, short-term profit, constructing cost	P, q	$P(d)=\begin{cases}0 & \text{if } d\le q\\ \dfrac{d-q}{p-q} & \text{if } q<d\le p\\ 1 & \text{if } d>p\end{cases}$

(continued)

Table 4.42 (continued)

Type	Description	Condition	Parameter	Preference function
	linearly in the interval (q < d ≤ p), the existence of the full priority of an alternative in interval (d > p)			
VI (Gaussian criterion)	Smooth transition based on normal distribution. Lack of difference in the interval (d ≤ 0), an increase in the priority rate of alternative in the interval (d > 0)	Appearance, quality, and safety	σ	$P(d)=\begin{cases}0 & d\leq 0\\ 1-e^{-\frac{d^2}{2\sigma^2}} & d>0\end{cases}$

preference function. It is recommended to use type 1, 2 and 4 functions for qualitative indicators and type 3, 5 and 6 functions for quantitative indicators. Additionally, the preference value of alternative A_l over A_h criterion X_j can be expressed as $P_{l,\ h,\ j}$.

Experts determine the preference function for each indicator along with the parameters q, p, and σ as needed, based on the selected preference function.

If a type 6 preference function is chosen for an indicator, experts first select the parameters p and q to estimate σ (standard deviation). Then, they determine the value of σ between the two thresholds p and q.

Step 3. Multiply the weight of each indicator by its corresponding matrix to determine the weighted preference value of alternative A_l over A_h:

$$\pi_{l,h,j} = P_{l,h,j} \times w_j.$$

Step 4. Calculate the aggregated weighted value matrix π. The elements of this matrix are obtained by summing the weighted preference values $\pi_{l,\ h,\ j}$ for each pair of alternatives (A_l, A_h) across all indicators:

$$\pi_{l,h} = \sum\nolimits_{j=1}^{n} \pi_{l,h,j}.$$

The following relations always hold for $\pi_{l,\ h}$:

$$\begin{cases}\pi_{l,h} = \pi_{h,h} = 0\\ 0 \leq \pi_{l,h} \leq 1\\ 0 \leq \pi_{l,h} + \pi_{h,l} \leq 1\end{cases}.$$

Step 5. Calculate the row average as the *positive flow* (Φ^+) and the column average as the *negative flow* (Φ^-) for all alternatives using the following relation. Note that since no alternative is compared with itself, the number of elements in each row and column is equal to *m*−*1*:

$$\Phi_l^+ = \frac{1}{m-1} \sum\nolimits_{\substack{h=l\\ h \neq l}}^{m} \pi_{l,h} \quad , \quad \Phi_l^- = \frac{1}{m-1} \sum\nolimits_{\substack{h=l\\ h \neq l}}^{m} \pi_{h,h} \; .$$

For simplicity, the term $\frac{1}{m-1}$ may be omitted, defining the positive and negative flows directly as the row sum and column sum, respectively.

Step 6. Based on the positive and negative flows, three cases emerge:

Case 1: Strict Preference

If any of the following three conditions is satisfied, then alternative A_l is strictly preferred to A_h (denoted as $A_l P^I A_h$):

$$\begin{cases} \Phi_l^- < \Phi_h^- \\ \Phi_l^+ < \Phi_h^+ \end{cases}, \begin{cases} \Phi_l^- < \Phi_h^- \\ \Phi_l^+ = \Phi_h^+ \end{cases}, \begin{cases} \Phi_l^- = \Phi_h^- \\ \Phi_l^+ > \Phi_h^+ \end{cases}.$$

In this case, the strong alternative A_l is compared against the weak alternative A_h. The positive and negative flows are consistent (both indicate the superiority of one alternative) and can be relied upon.

Case 2: Indifference

If the conditions $\Phi_l^- = \Phi_h^-$ and $\Phi_l^+ = \Phi_h^+$ hold, in other words, when the negative flows are equal to each other and the positive flows are equal to each other, then alternatives A_l and A_h are indifferent ($A_l I^I A_h$).

Case 3: Incomparability

If either of the following two conditions occurs, alternatives A_l and A_h cannot be compared ($A_l R^I A_h$):

$$\begin{cases} \Phi_l^- > \Phi_h^- \\ \Phi_l^+ > \Phi_h^+ \end{cases}, \begin{cases} \Phi_l^- < \Phi_h^- \\ \Phi_l^+ < \Phi_h^+ \end{cases}.$$

In this scenario, the strengths of each alternative are contrasted with the weaknesses of the other. This occurs when A_l outperforms A_h on some criteria, while A_h outperforms A_l on others. Consequently, the positive and negative flows are inconsistent. In such cases, it is reasonable to conclude that the alternatives are incomparable.

The relations I^I, P^I, and R^I represent standard pairwise preference relations, I (indifference), P (strict preference), and R (incomparability).

The superscript I indicates that these relations are derived from the PROMETHEE methodology. To avoid confusion, relations obtained from PROMETHEE I are denoted with superscript I (e.g., P^I, I^I, and R^I), and the relations derived from PROMETHEE II are denoted as P^{II}, I^{II}, and R^{II}. Relations from other PROMETHEE techniques use different superscripts.

4.10.2.2 The PROMETHEE II Technique

PROMETHEE II represents an extended version of PROMETHEE I for complete ranking of alternatives. This technique was introduced concurrently with PROMETHEE I in the same seminal paper. Both methods follow identical steps through Step 5, with their divergence beginning at Step 6.

Step 6: Net flow calculation. Calculate the net flow of alternative A_l using the following relation:

$$\Phi_l = \Phi_l^+ - \Phi_l^- .$$

The following property always holds for net flow values:

$$-1 \le \Phi_l \le 1, \ \sum_{i=1}^{m} \Phi_l = 0.$$

Step 7: Rank the alternatives in descending order of Φ_l. Note that when comparing alternatives A_l and A_h based on their net flows, two cases may arise:

Case 1: Strict preference. Alternative A_l is strictly preferred to A_h (denoted $A_l P^{II} A_h$) if $\Phi_l > \Phi_h$.

Case 2: Indifference. Alternatives A_l and A_h are indifferent (denoted $A_l I^{II} A_h$) if $\Phi_l = \Phi_h$.

In the PROMETHEE II method, all alternatives are comparable, and no incomparability exists. The relations obtained from PROMETHEE I are referred to as a *partial ranking (type 1)*, while those from PROMETHEE II constitute a *complete ranking (type 2)*.

Brans recommends using both PROMETHEE I and PROMETHEE II for real-world problems. The complete ranking (PROMETHEE II) is easier to interpret, whereas the partial ranking (PROMETHEE I) only displays *confirmed superiorities*.

Furthermore, the incomparability analysis—unique to PROMETHEE I—leads to more comprehensive conclusions by identifying cases where alternatives cannot be definitively ranked.

4.10.2.3 The PROMETHEE III Technique

In the PROMETHEE I technique, it was established that if for a pair of alternatives, we simultaneously have $\Phi_l^- = \Phi_h^-$ and $\Phi_l^+ = \Phi_h^+$, then the alternatives are indifferent. Similarly, in PROMETHEE II, the indifference condition is expressed as $\Phi_l = \Phi_h$.

Since preference function values (and consequently, positive, negative, and net flows) are typically real numbers, strict indifference conditions rarely occur. Moreover, two alternatives with different but closely valued net flows might be considered indifferent by the decision-maker (DM), suggesting that indifference and preference thresholds should also be applied to flows.

To address these limitations, Brans et al. (1986) proposed the PROMETHEE III technique. The methodology remains identical to PROMETHEE II through Step 6.

Step 7: For each alternative, define an indifference interval using the following relation:

$$\begin{cases} L_l = \overline{\Phi}_l - \alpha S_l \\ U_l = \overline{\Phi}_l + \alpha S_l \end{cases}$$

where L_l and U_l represent the lower and upper bounds of alternative A_l's indifference interval, respectively.

When two decision-makers (DMs) differ in their assessments, the probability of indifference between alternatives can be modeled using a coefficient $\alpha \in [0, 1]$. Maximum indifference probability occurs at $\alpha = 1$ and PROMETHEE II equivalence is achieved at $\alpha = 0$ (strict ranking). The value $\alpha = 0.15$ is typically recommended for detecting significant differences between alternatives (Safari et al., 2024).

The standard deviation s_l of the net flow for alternative l (Φ_l) is derived from:

$$s_l = \sqrt{\frac{1}{m} \sum_{h=1}^{m} \left(\pi_{l,h} - \pi_{h,l} - \overline{\Phi}_l\right)^2}$$

where

$\pi_{l,\ h}$: Preference of A_l over A_h.

$\overline{\Phi}_l = \frac{\sum \Phi_l}{m}$: Mean net flow of A_l.

Step 8. Compare the indifference intervals of alternative pairs. Based on the derived intervals, two distinct cases may emerge:

Case 1: Strict Preference

If $L_l > U_h$ holds true or, equivalently, when the indifference intervals of these two alternatives are disjoint ($I_l \cap I_h = \varnothing$) and furthermore the interval corresponding to A_l is positioned entirely to the right of the interval corresponding to A_h, then alternative A_l is strictly preferred to A_h.

Case 2: Indifference

Occurs when either of the upper bound of one alternative exceeds the lower bound of the other ($U_h > L_l$and $U_l > L_h$) or one interval fully contains the other ($L_h \le L_1$and $U_h \ge U_l$) or ($L_l \le L_h$and $U_l \ge U_h$).

Note on Transitivity in PROMETHEE III The transitivity property holds for strict preference relations (P^{III}), but not necessarily for indifference relations (I^{III}) its application in indifference cases depends on the decision-maker's (DM) judgment.

Example 4.4
A company intends to purchase a number of vehicles. The vehicles available for purchase by the company are as follows. Given the values of the options (vehicles) based on the five criteria and the specified preference thresholds as detailed in the tables below, if a bipolar Likert scale is used for quantification, determine the relationships between the vehicles using the PROMETHEE I technique.

Criteria (Indices)

- X_1^+: *Maximum fuel efficiency*
- X_2^+: *Structural strength*
- X_3^- : *Acceleration (0–100 km/h)*
- X_4^+: *Body design*
- X_5^- : *Safety*

Alternatives (Options)

1. ***A₁***: *BMW M4 E4Y*
2. ***A₂***: *Mercedes-Benz C-Class*
3. ***A₃***: *Audi R5†*
4. ***A₄***: *Ford Mustang GT*
5. ***A₅***: *Porsche 911 Carrera*

Preference function type (P_j)
Indifference threshold (q_j)
Preference threshold (p_j)
Standard deviation (σ_j)

Performance Ratings

Alternative	X_1^+	X_2^+	X_3^-	X_4^+	X_5^-
A_1	Low	Extremely low	4.30	312	9
A_2	Medium	Low	4.45	307	12
A_3	Extremely low	High	5.21	280	7.5

(continued)

Alternative	X_1^+	X_2^+	X_3^-	X_4^+	X_5^-
A_4	Slightly low	Slightly low	3.94	241	14
A_5	Low	Slightly low	4.70	284	10.5
W_j	0.26	0.19	0.18	0.22	0.15

Preference Function Parameters

X_j	P_j	q_j	P_j	σ_j
X_1	Type 2 (quasi-criterion)	1	–	–
X_2	Type 4 (linear)	1	2	–
X_3	Type 3 (V-shaped)	–	0.8	–
X_4	Type 6 (Gaussian)	–	–	30
X_5	Type 5 (level)	3	5	–

Solution

To begin the analysis, we first quantify the criteria and present the performance evaluation matrix for the alternatives.

Alternative	X_1	X_2	X_3	X_4	X_5
A_1	3	1	4.30	312	9
A_2	5	3	4.45	307	12
A_3	1	2	5.21	280	7.5
A_4	4	4	3.94	241	14
A_5	3	4	4.70	284	10.5
W_j	0.26	0.19	0.18	0.22	0.15

Step 1: Calculating pairwise differences in criterion values

X_1	A_1	A_2	A_3	A_4	A_5
A_1	–	−2	2	−1	0
A_2	2	–	4	1	2
A_3	−2	−4	–	−3	−2
A_4	1	−1	3	–	1
A_5	0	−2	2	−1	–

X_2	A_1	A_2	A_3	A_4	A_5
A_1	–	−2	−1	−3	−3
A_2	2	–	1	−1	−1
A_3	1	−1	–	−2	−2
A_4	3	1	2	–	0
A_5	3	1	2	0	–

X_3	A_1	A_2	A_3	A_4	A_5
A_1	–	−0.15	−0.91	0.36	−0.40
A_2	0.15	–	−0.76	0.51	−0.25
A_3	0.91	0.76	–	1.27	0.51
A_4	−0.36	−0.51	−1.27	–	−0.76

(continued)

X_3	A_1	A_2	A_3	A_4	A_5
A_5	0.40	0.25	−0.51	0.76	–

X_4	A_1	A_2	A_3	A_4	A_5
A_1	–	5	32	71	28
A_2	−5	–	27	66	23
A_3	−32	−27	–	39	−4
A_4	−71	−66	−39	–	−43
A_5	−28	−23	4	43	–

X_5	A_1	A_2	A_3	A_4	A_5
A_1	–	−3	1.5	−5	−1.5
A_2	3	–	4.5	−2	1.5
A_3	−1.5	−4.5	–	−6.5	−3
A_4	5	2	6.5	–	3.5
A_5	1.5	−1.5	3	−3.5	–

The pairwise difference matrices are antisymmetric (skew-symmetric), meaning:

$$d_j(A_l, A_h) = -d_j(A_h, A_l).$$

Thus, only the upper or lower triangular part needs explicit calculation.

Step 2: Calculating preference function values for pairwise differences

X_1	A_1	A_2	A_3	A_4	A_5
A_1	–	0	1	0	0
A_2	1	–	1	0	1
A_3	0	0	–	0	0
A_4	0	0	1	–	0
A_5	0	0	1	0	–

X_2	A_1	A_2	A_3	A_4	A_5
A_1	–	0	0	0	0
A_2	0.5	–	0	0	0
A_3	0	0	–	0	0
A_4	1	0	0.5	–	0
A_5	1	0	0.5	0	–

X_3	A_1	A_2	A_3	A_4	A_5
A_1	–	0.1875	1	0	0.5
A_2	0	–	0.95	0	0.3125
A_3	0	0	–	0	0
A_4	0.45	0.6375	1	–	0.95
A_5	0	0	0.6375	0	–

X_4	A_1	A_2	A_3	A_4	A_5
A_1	–	0.0138	0.4338	0.9392	0.3531

(continued)

X_4	A_1	A_2	A_3	A_4	A_5
A_2	0	–	0.3330	0.9111	0.2546
A_3	0	0	–	0.5704	0
A_4	0	0	0	–	0
A_5	0	0	0.6375	0	–

X_5	A_1	A_2	A_3	A_4	A_5
A_1	–	0	0	1	0
A_2	0	–	0	0	0
A_3	0	0.75	–	1	0
A_4	0	0	0	–	0
A_5	0	0	0	0.25	–

Step 3: Calculating weighted preference criteria

X_1	A_1	A_2	A_3	A_4	A_5
A_1	–	0	0.26	0	0
A_2	0.26	–	0.26	0	1
A_3	0	0	–	0	0
A_4	0	0	0.26	–	0
A_5	0	0	0.26	0	–

X_2	A_1	A_2	A_3	A_4	A_5
A_1	–	0	0	0	0
A_2	0.095	–	0	0	0
A_3	0	0	–	0	0
A_4	0.19	0	0.095	–	0
A_5	0.19	0	0.095	0	–

X_3	A_1	A_2	A_3	A_4	A_5
A_1	–	0.0338	0.18	0	0.09
A_2	0	–	0.171	0	0.0562
A_3	0	0	–	0	0
A_4	0.081	0.1147	0.18	–	0.171
A_5	0	0	0.1147	0	–

X_4	A_1	A_2	A_3	A_4	A_5
A_1	–	0.003	0.0954	0.2066	0.0777
A_2	0	–	0.0733	0.2004	0.056
A_3	0	0	–	0.1255	0
A_4	0	0	0	–	0
A_5	0	0	0.19	0.1412	–

X_5	A_1	A_2	A_3	A_4	A_5
A_1	–	0	0	0.15	0
A_2	0	–	0	0	0
A_3	0	0.1125	–	0.15	0
A_4	0	0	0	–	0
A_5	0	0	0	0.0375	–

Step 4: Calculation of the weighted preference matrix (π)

π	A_1	A_2	A_3	A_4	A_5
A_1	–	0.0368	0.5354	0.3566	0.1677
A_2	0.3550	–	0.5043	0.2004	0.3722
A_3	0	0.1125	–	0.2755	0
A_4	0.271	0.1147	0.535	–	0.171
A_5	0.19	0	0.4716	0.1787	–

Step 5: Calculation of positive and negative outranking flows

Alternatives	Φ^+	Φ^-
A_1	0.2741	0.204
A_2	0.358	0.066
A_3	0.097	0.5116
A_4	0.2729	0.2528
A_5	0.2101	0.1777

Step 6: Establishing outranking relationships between alternatives

	A_1	A_2	A_3	A_4	A_5
A_1	–		P^I	P^I	R^I
A_2	P^I	–	P^I	P^I	P^I
A_3			–		
A_4			P^I	–	R^I
A_5	R^I		P^I	R^I	–

Note that PROMETHEE I presents pairwise preference relations between alternatives. To obtain a complete ranking of alternatives, the decision-maker should utilize the PROMETHEE II method, implemented as follows:

Alternatives	Φ^+	Φ^-	$\Phi_l = \Phi_l^+ - \Phi_l^-$	Rank
A_1	0.2741	0.204	0.0701	2
A_2	0.358	0.066	0.292	1
A_3	0.097	0.5116	−0.4146	5
A_4	0.2729	0.2528	0.0201	4
A_5	0.2101	0.1777	0.0324	3

4.10.2.4 Case Study

We shall analyze the PROMETHEE II implementation in the case study of Safari et al. (2024) (Table 4.43)**.**

Step 1: Calculation of preference function values for pairwise differences (Tables 4.44, 4.45, 4.46, 4.47, 4.48, 4.49, 4.50 and 4.51)

Step 2: Calculation of the weighted preference matrix (π) (Table 4.52)

Table 4.43 Aggregated decision matrix

	C_1			C_2			C_3			C_4			C_5			C_6			C_7			C_8		
A_1	0.6	0.7	0.8	0.2	0.3	0.4	0.3	0.4	0.5	0.4	0.4	0.6	0.4	0.5	0.6	0.5	0.6	0.7	0.4	0.5	0.7	0.2	0.3	0.5
A_2	0.2	0.3	0.5	0.3	0.5	0.6	0.1	0.2	0.4	0.3	0.4	0.6	0.3	0.4	0.5	0.3	0.4	0.6	0.3	0.4	0.5	0.1	0.2	0.4
A_3	0.7	0.8	0.9	0.6	0.7	0.9	0.7	0.8	0.9	0.5	0.6	0.7	0.6	0.7	0.8	0.6	0.7	0.8	0.6	0.7	0.8	0.6	0.7	0.8
A_4	0.5	0.6	0.7	0.2	0.3	0.4	0.4	0.5	0.6	0.4	0.5	0.7	0.2	0.3	0.5	0.3	0.4	0.6	0.4	0.5	0.6	0.5	0.6	0.7
A_5	0.3	0.4	0.5	0.5	0.6	0.7	0.4	0.5	0.7	0.3	0.4	0.6	0.4	0.5	0.6	0.3	0.5	0.6	0.4	0.5	0.6	0.6	0.7	0.8
A_6	0.4	0.5	0.6	0.4	0.5	0.7	0.5	0.6	0.7	0.3	0.4	0.5	0.2	0.3	0.4	0.4	0.5	0.6	0.2	0.4	0.5	0.4	0.5	0.7
W_j	0.2	0.2	0.4	0.2	0.3	0.4	0.3	0.4	0.6	0.5	0.6	0.7	0.5	0.6	0.7	0.4	0.5	0.6	0.5	0.6	0.7	0.4	0.5	0.6

Table 4.44 Preference function values for criterion 1

C_1	A_1			A_2			A_3			A_4			A_5			A_6		
A_1	0	0	0	0	0	1	0	0	0	0	0	0	0	0	0	0	0	0
A_2	0	0	0	0	0	0	0	0	0	0	0	0	0	0	0	0	0	0
A_3	0	0	0	0	0	1	0	0	0	0	0	0	0	0	1	0	0	1
A_4	0	0	0	0	0	1	0	0	0	0	0	0	0	0	0	0	0	0
A_5	0	0	0	0	0	0	0	0	0	0	0	0	0	0	0	0	0	0
A_6	0	0	0	0	0	0	0	0	0	0	0	0	0	0	0	0	0	0

Table 4.45 Preference function values for criterion 2

C_2	A_1			A_2			A_3			A_4			A_5			A_6		
A_1	0	0	0	0	0	0	0	0	0	0	0	0	0	0	0	0	0	0
A_2	−0	0	0.4	0	0	0	0	0	0	0	0	0	0	0	0	0	0	0
A_3	0.4	1	0.8	0	0	0.6	0	0	0	0	0.4	1	−0	0.2	0.5	0.2	0	0.7
A_4	−0	0	0.4	−0	0	0.2	0	0	0	0	0	0	0	0	0	−0	0	0.3
A_5	0.1	0	0.6	0	0	0.5	0	0	0	−0	0.2	0	0	0	0	−0	0	0.5
A_6	−0	0	0.4	−0	0	0.3	0	0	0	0	0	0	0	0	0	0	0	0

Table 4.46 Preference function values for criterion 3

C_3	A_1			A_2			A_3			A_4			A_5			A_6		
A_1	0	0	0	0	0	0	0	0	0	0	0	0	0	0	0	0	0	0
A_2	−0	0	0.4	0	0	0	0	0	0	−0	0.1	0	0	0	0	0	0	0
A_3	0.2	0	0.7	0	0	0.5	0	0	0	0	0.4	1	−0	0.2	0.4	−0	0	0.5
A_4	−0	0	0.2	0	0	0	0	0	0	0	0	0	0	0	0	0	0	0
A_5	0.1	0	0.5	0	0	0.4	0	0	0	0	0.3	0	0	0	0	−0	0	0.3
A_6	−0	0	0.5	−0	0	0.3	0	0	0	−0	0.2	0	0	0	0	0	0	0

Table 4.47 Preference function values for criterion 4

C_4	A_1			A_2			A_3			A_4			A_5			A_6		
A_1	0	0	0	−0	0	0.3	0	0	0	0	0	0	−0	0	0.3	−0	0	0.3
A_2	0	0	0	0	0	0	0	0	0	0	0	0	−0	0	0.3	−0	0	0.3
A_3	−0	0	0.3	−0	0	0.4	0	0	0	−0	0.1	0	−0	0.2	0.4	−0	0	0.4
A_4	−0	0	0.3	−0	0	0.3	0	0	0	0	0	0	−0	0.1	0.4	−0	0	0.4
A_5	0	0	0	0	0	0	0	0	0	0	0	0	0	0	0	−0	0	0.3
A_6	0	0	0	0	0	0	0	0	0	0	0	0	0	0	0	0	0	0

Table 4.48 Preference function values for criterion 5

C_5	A_1			A_2			A_3			A_4			A_5			A_6		
A_1	0	0	0	−0	0	0.3	0	0	0	−0	0.2	0	−0	0	0.2	−0	0	0.4
A_2	0	0	0	0	0	0	0	0	0	−0	0.1	0	0	0	0	−0	0	0.3
A_3	−0	0	0.4	−0	0	0.5	0	0	0	0	0.4	1	−0	0.2	0.4	0.1	0	0.6
A_4	0	0	0	0	0	0	0	0	0	0	0	0	0	0	0	−0	0	0.2
A_5	0	0	0	−0	0	0.3	0	0	0	−0	0.2	0	0	0	0	−0	0	0.4
A_6	0	0	0	0	0	0	0	0	0	0	0	0	0	0	0	0	0	0

Table 4.49 Preference function values for criterion 6

C_6	A_1			A_2			A_3			A_4			A_5			A_6		
A_1	0	0	0	−0	0	0.4	0	0	0	−0	0.1	0	−0	0.1	0.3	−0	0	0.3
A_2	0	0	0	0	0	0	0	0	0	0	0	0	0	0	0	0	0	0
A_3	−0	0	0.3	−0	0	0.5	0	0	0	0	0.2	0	0	0.2	0.4	−0	0	0.4
A_4	0	0	0	−0	0	0.3	0	0	0	0	0	0	0	0	0	0	0	0
A_5	0	0	0	−0	0	0.3	0	0	0	−0	0	0	0	0	0	0	0	0
A_6	0	0	0	−0	0	0.3	0	0	0	−0	0.1	0	−0	0.1	0.3	0	0	0

Table 4.50 Preference function values for criterion 7

C_7	A_1			A_2			A_3			A_4			A_5			A_6		
A_1	0	0	0	−0	0	0.4	0	0	0	−0	0	0	−0	0	0.3	−0	0	0.4
A_2	0	0	0	0	0	0	0	0	0	0	0	0	0	0	0	−0	0	0.3
A_3	−0	0	0.4	−0	0	0.5	0	0	0	−0	0.2	0	−0	0.2	0.4	0.1	0	0.6
A_4	0	0	0	−0	0	0.3	0	0	0	0	0	0	−0	0	0.2	−0	0	0.4
A_5	0	0	0	−0	0	0.3	0	0	0	0	0	0	0	0	0	−0	0	0.4
A_6	0	0	0	0	0	0	0	0	0	0	0	0	0	0	0	0	0	0

Table 4.51 Preference function values for criterion 8

C_8	A_1			A_2			A_3			A_4			A_5			A_6		
A_1	0	0	0	−0	0	0.3	0	0	0	0	0	0	0	0	0	0	0	0
A_2	0	0	0	0	0	0	0	0	0	0	0	0	0	0	0	0	0	0
A_3	0.2	0	0.6	0	0	0.7	0	0	0	−0	0.2	0	−0	0.1	0.3	−0	0	0.4
A_4	−0	0	0.4	−0	0	0.5	0	0	0	0	0	0	0	0	0	−0	0	0.2
A_5	0.1	0	0.5	0	0	0.6	0	0	0	−0	0.1	0	0	0	0	−0	0	0.3
A_6	−0	0	0.4	−0	0	0.5	0	0	0	0	0	0	0	0	0	0	0	0

Table 4.52 Weighted preference matrix (**π**)

π	A₁			A₂			A₃			A₄			A₅			A₆		
A_1	0.00	0.00	0.00	−0.49	0.31	1.26	0.00	0.00	0.00	−0.18	0.19	0.73	−0.30	0.16	0.92	−0.20	0.30	1.09
A_2	−0.04	0.12	0.40	0.00	0.00	0.00	0.00	0.00	0.00	−0.11	0.12	0.41	−0.11	0.01	0.19	−0.27	0.11	0.62
A_3	0.07	0.95	2.13	0.08	1.14	2.47	0.00	0.00	0.00	0.03	0.89	2.10	−0.12	0.70	1.93	0.09	0.99	2.29
A_4	−0.13	0.25	0.78	−0.36	0.36	1.26	0.00	0.00	0.00	0.00	0.00	0.00	−0.17	0.11	0.57	−0.34	0.21	1.10
A_5	0.05	0.40	0.87	−0.36	0.45	1.50	0.00	0.00	0.00	−0.19	0.30	1.02	0.00	0.00	0.00	−0.33	0.32	1.29
A_6	−0.02	0.26	0.68	−0.20	0.26	0.93	0.00	0.00	0.00	−0.08	0.11	0.41	−0.10	0.03	0.27	0.00	0.00	0.00

Table 4.53 Positive and negative outranking flows and net flows

π	Φ^{+}			Φ^{-}			$\Phi_l = \Phi_l^{+} - \Phi_l^{-}$		
A_1	−1.17	0.96	4.00	−0.07	1.98	4.85	−6.02	−1.01	4.06
A_2	−0.52	0.36	1.62	−1.33	2.52	7.42	−7.93	−2.17	2.95
A_3	0.14	4.66	10.91	0.00	0.00	0.00	0.14	4.66	10.91
A_4	−1.00	0.93	3.72	−0.53	1.61	4.68	−5.68	−0.68	4.24
A_5	−0.82	1.48	4.68	−0.81	1.00	3.87	−4.70	0.48	5.48
A_6	−0.40	0.66	2.29	−1.04	1.93	6.40	−6.80	−1.28	3.33

Table 4.54 Comparative analysis and ranking alternatives

	A_1	A_2	A_3	A_4	A_5	A_6
A_1	1.00	0.89	1.00	1.00	1.00	0.97
A_2	1.00	1.00	1.00	1.00	1.00	1.00
A_3	0.41	0.29	1.00	0.43	0.56	0.35
A_4	0.97	0.85	1.00	1.00	1.00	0.94
A_5	0.85	0.74	1.00	0.88	1.00	0.82
A_6	1.00	0.92	1.00	1.00	1.00	1.00
V(Ci>C1,C2,C3,C4,C5,C6)	0.41	0.29	1.00	0.43	0.56	0.35
Weight	0.13	0.10	0.33	0.14	0.18	0.11
Rank	4	6	1	3	2	5

Step 3: Calculation of positive and negative outranking flows, net flows, and ranking of alternatives (Table 4.53)

Step 4: Comparative analysis of alternatives A_l and A_h based on net flows and ranking alternatives (Table 4.54)

References

Atanassov, K. T. (1986). Intuitionistic fuzzy sets. *Fuzzy Sets and Systems, 20*, 87–96. https://doi.org/10.1016/S0165-0114(86)80034-3

Brans, J. P., Vincke, Ph., & Mareschal, B. (1986). How to select and how to rank projects: The Promethee method. *European Journal of Operational Research, 24*(2), 228–238.

Chen, S. J., & Hwang, C. L. (1992). *Fuzzy multiple attribute decision making*. Springer-Verlag.

Homayounfar, M., Goudarzvand Chegini, M., & Daneshvar, A. (2018). Prioritization of green supply chain suppliers using a hybrid fuzzy multi criteria decision making approach. *Journal of Operational Research and Its Applications, 15*(2), 41–61.

Niu, D., Song, Z., Wang, M., & Xiao, X. (2017). Improved TOPSIS method for power distribution network investment decision-making based on benefit evaluation indicator system. *International Journal of Energy Sector Management, 11*, 595–608.

Roy, B. (1968). Classement et choix en présence de points de vue multiples (la méthode ELECTRE). *Revue française d'informatique et de recherche opérationnelle (R.I.R.O.), 2*(8), 57–75.

Roy, B. (1978). ELECTRE III: Un algorithme de classements fondé sur une repré- sentation floue des préférences en présence de critères multiples. *Cahiers du Centre d'études de recherche operationnelle, 20*, 3–24.

Roy, B., & Bertier, P. (1972). La méthode ELECTRE II - Une application au média-planning. In M. Ross (Ed.), *OR'72: Proceedings of the Sixth International Conference on Operational Research* (pp. 291–302). Amsterdam: North-Holland Publishing Company.

Safari, H., Soufi, M., & Aghasi, E. (2014). SELECT A HYPERMARKET LOCATION BASED ON FUZZY MULTI CRITERIA DECISION MAKING (F-MCDM) TECHNIQUES (HYBRID OF F-DELPHI, F-AHP, F-LLSM and F-PROMETHEE). *Kuwait Chapter of Arabian Journal of Business and Management Review, 4*(1), 1–20.

Safari, S., Soufi, M., & Amirteimoori, A. (2024). Assessment of internal and external consistency of intuitive decision-making methods in comparison with three multi-criteria decision-making methods. *Journal of Operational Research and Its Applications, 2*(81), 141–157.

Shih, H.-S., Shyur, H.-J., & Lee, E. S. (2007). An extension of TOPSIS for group decision making. *Mathematical and Computer Modelling, 45*(7–8), 801–813. https://doi.org/10.1016/j.mcm.2006.03.023

Smarandache, F. (1999). *A unifying field in logics. Neutrosophy: Neutrosophic probability, set and logic*. American Research Press.

Torra, V. (2010). Hesitant fuzzy sets. *International Journal of Intelligent Systems, 25*, 529–539. https://doi.org/10.1002/int.20418

Torra, V., & Narukawa, Y. (2009). On Hesitant Fuzzy Sets and Decision. *The 18th IEEE International Conference on Fuzzy Systems* (pp. 1378–1382). Jeju Island. https://doi.org/10.1109/FUZZY.2009.5276884

Wang, H., Smarandache, F., Zhang, Y., & Sunderraman, R. (2005). Interval neutrosophic sets and logic: Theory and applications in computing. *Neutrosophic Book Series, 5*.

Chapter 5
Multi-criteria Decision-Making Problems in Excel

Learning Objectives

Upon completion of this chapter, the reader will be able to:

1. **Implement** fundamental multi-criteria decision-making (MCDM) techniques, including SAW (simple additive weighting), WPM (weighted product model), and WASPAS (weighted aggregated sum product assessment), using Microsoft Excel's built-in functions for data normalization and aggregation.
2. **Apply** the MAUT (multi-attribute utility theory) method in Excel to model decision-maker risk preferences (risk-averse, risk-seeking, risk-neutral) through different utility functions for each criterion.
3. **Execute** the COPRAS (complex proportional assessment) technique within the Excel environment to evaluate alternatives by simultaneously considering beneficial and non-beneficial criteria.
4. **Construct** and manipulate decision matrices, perform normalization procedures specific to maximization and minimization criteria, and calculate weighted scores for systematic alternative evaluation.
5. **Develop** automated Excel templates for MCDM problems by effectively utilizing functions such as SUMPRODUCT, MAX, MIN, RANK, and PRODUCT to streamline the decision-analysis process.
6. **Compare** the results and practical implementation nuances of different MCDM methods when applied to the same dataset within a spreadsheet framework.

A. Amirteimoori, M. Soufi, *Industrial Decision Analysis*, Multiple Criteria Decision Making, https://doi.org/10.1007/978-3-032-24299-0_5

5.1 Introduction

In the age of information and data-driven decision-making, the application of systematic and structured methods for solving complex multi-criteria problems has become an undeniable necessity. However, one of the fundamental challenges faced by researchers and managers in organizations is the lack of access to specialized and costly multi-criteria decision-making software such as Expert Choice, Super Decisions, and LINGO. This limitation, particularly in small- and medium-sized enterprises, universities, and research institutions with constrained budgets, can pose a serious obstacle to implementing scientific decision-making methodologies.

Microsoft Excel, as one of the most accessible and widely used analytical tools, offers an efficient and cost-effective solution to overcome this challenge. Equipped with advanced computational capabilities, diverse functions, data analysis tools, and VBA programming environment, this software enables the implementation of a broad spectrum of multi-criteria decision-making techniques with acceptable accuracy and speed. Furthermore, managers and analysts' widespread familiarity with the Excel user interface minimizes the learning curve and facilitates seamless exchange of results across different organizational teams.

This chapter is dedicated to the practical application of Microsoft Excel in implementing multi-criteria decision-making techniques. In this chapter, six widely used methods including simple additive weighting (SAW), weighted product model (WPM), weighted aggregated sum product assessment (WASPAS), multiple attribute utility theory (MAUT), and complex proportional assessment (COPRAS) are elaborated through a step-by-step, hands-on approach.

For each of these methods, alongside the presentation of theoretical foundations and mathematical formulations, complete implementation in the Excel environment is demonstrated using basic functions, array functions, and data analysis tools. Numerous numerical examples and real-world case studies assist readers in developing a deep understanding of the computational process and result interpretation. Additionally, ready-made templates designed for each method enable reuse and customization for similar problems.

The ultimate objective of this chapter is to empower managers, practitioners, and researchers to independently implement multi-criteria decision-making methods in the familiar and accessible Excel environment, thereby minimizing dependence on costly specialized software while accelerating and facilitating the decision-making process within organizations.

5.2 The SAW Technique (Simple Additive Weighting Method)

The simple additive weighting (SAW) method, also known as the weighted sum method, represents the most fundamental approach in multi-attribute decision-making (MADM). This technique was originally developed by Hwang and Yoon in 1981. Due to its computational simplicity and intuitive application, SAW has become one of the most prevalent and widely adopted methods in MADM problems.

The SAW method can be considered the most straightforward and direct approach for addressing multi-criteria decision-making challenges. This is primarily because it employs a linear additive utility function to aggregate decision-makers' preferences across various criteria. However, it's crucial to note that this technique maintains validity only under the fundamental assumption that the decision criteria are mutually independent or noninteracting.

From a methodological perspective, SAW operates through a systematic normalization process followed by weighted aggregation, making it particularly suitable for decision scenarios where:

- Criteria weights can be reliably estimated.
- Performance measures are quantitatively comparable.
- The independence assumption of criteria holds reasonably true.

The method's popularity stems from its:

- Computational efficiency
- Ease of implementation
- Transparent decision process
- Minimal data requirements

Nevertheless, practitioners should be aware of its limitations when dealing with:

- Highly correlated criteria
- Nonlinear preference structures
- Qualitative performance measures requiring special normalization techniques

Steps of the Simple Additive Weighting (SAW) Method

In a multi-criteria decision-making problem with n criteria and m alternatives, the procedure for selecting the optimal alternative using the SAW method is as follows:

1. **Construct the decision matrix.**
2. **Normalize the decision matrix.**
3. **For maximization criteria (positive):**

$$r_{ij} = \frac{x_{ij}}{Max\, x_j}.$$

4. **For minimization criteria (negative):**

Table 5.1 Decision matrix

	A	B	C	D	E	F
1		C_1^-	C_2^+	C_3^-	C_4^-	C_5^+
2	A_1	24	750	1	40	2
3	A_2	46	1160	4	20	2
4	A_3	30	1430	5	40	5
5	A_4	35	1020	3	5	3
6	W_j	0.35	0.2	0.1	0.2	0.15
7	Upper bound	46.00	1430.00	5.00	40.00	5.00
8	Lower bound	24.00	750.00	1.00	5.00	2.00

$$r_{ij} = \frac{Min\, x_j}{x_{ij}}.$$

5. **Construct the weighted matrix.**
6. **Select the optimal alternative.**
7. **Rank alternatives.**

Example 5.1 A contracting company has identified four potential sites for the construction of an industrial warehouse. The relevant data is presented in the table below. Using the SAW (simple additive weighting) method, determine the optimal location (Table 5.1).

For data normalization, the upper and lower bounds of the indicators are determined, as specified in rows 7 and 8.

To calculate the upper bound in Excel, enter the following formula in the target cell:

$$= \text{MAX (B2 : B6)}.$$

For lower bound (minimum): =MIN (B2:B6).

Since the first criterion is a cost-based (negative) indicator, to normalize cell B10, we divide the lower bound of criterion 1 (from cell B8 = 24) by the value in cell B2 (24). To drag and drop this formula across the entire column while keeping the reference to B8 fixed (absolute reference), we add a dollar sign before the row and column identifiers like this: B$8 (Tables 5.2 and 5.3).

Finally, the total score for each alternative is calculated by summing the weighted normalized scores of all its criteria, which is then used to rank the alternatives.

To perform the ranking in cell C20, enter the following formula and extend it down to cell C23 (Table 5.4):

$$= \text{RANK(B20, \$B\$20 : \$B\$23, 0)}.$$

Table 5.2 Normalized decision matrix

	A	B	C	D	E	F
9		C_1^-	C_2^+	C_3^-	C_4^-	C_5^+
10	A_1	=$B$8/B2 (1.0000)	=C2/$C$8 (0.5245)	=$D$8/D2 (1.0000)	=$E$8/E2 (0.1250)	F2/$F$8 (0.4000)
11	A_2	=$B$8/B3 (0.5217)	0.8112	0.2500	0.2500	0.4000
12	A_3	0.8000	1.0000	0.2000	0.1250	1.0000
13	A_4	0.6857	0.7133	0.3333	1.0000	0.6000

Table 5.3 Weighted normalized decision matrix

	A	B	C	D	E	F
14		C_1^-	C_2^+	C_3^-	C_4^-	C_5^+
15	A_1	=B10*$B$6 (0.3500)	=C10*$C$6 (0.1049)	=D10*$D$6 (0.1000)	=E10*$E$6 (0.0250)	=F10*$F$6 0.0600
16	A_2	0.1826	0.1622	0.0250	0.0500	0.0600
17	A_3	0.2800	0.2000	0.0200	0.0250	0.1500
18	A_4	0.2400	0.1427	0.0333	0.2000	0.0900

Table 5.4 Weighted alternative ranking

	A	B	C
19		SUM	RANK
20	A_1	=SUM(B15:F15) 0.6399	=RANK(B20,$B$20:$B$23,0) 3
21	A_2	=SUM(B16:F16) 0.4798	4
22	A_3	=SUM(B17:F17) 0.6750	2
23	A_4	=SUM(B18:F18) 0.7060	1

5.3 The Weighted Product Model (WPM)

The weighted product model (WPM) and weighted sum model (WSM or SAW) are methods in multi-criteria decision-making (MCDM) used for prioritizing alternatives based on various criteria and attributes. Zavadskas combined the results of these two methods and introduced the WASPAS (weighted aggregated sum product assessment) method.

The weighted product model (WPM) is one of the multi-criteria decision-making (MCDM) methods used to select the best alternative. This method was introduced by Bridgman (1922) and Miller and Starr (1960, 1969). The weighted product model is designed based on the well-known weighted sum model (WSM).

Table 5.5 Decision matrix

	A	B	C	D	E	F
1		C_1^+	C_2^-	C_3^-	C_4^+	C_5^-
2	A_1	4	1091	20	17700	5
3	A_2	9	735	30	16500	9
4	A_3	3	1082	20	23800	5
5	A_4	4	977	10	16200	8
6	A_5	7	856	35	28400	4
7	W_j	0.27	0.13	0.25	0.16	0.19

Table 5.6 Normalized decision matrix

	A	B	C	D	E	F
8		C_1^+	C_2^-	C_3^-	C_4^+	C_5^-
9	A1	=B2/MAX(B$2:B$6) 0.4444	=MIN(C$2:C$6)/C2 0.6737	0.5000	0.6232	0.8000
10	A2	1.0000	1.0000	0.3333	0.5810	0.4444
11	A3	0.3333	0.6793	0.5000	0.8380	0.8000
12	A4	0.4444	0.7523	1.0000	0.5704	0.5000
13	A5	0.7778	0.8586	0.2857	1.0000	1.0000

The weighted sum model (WSM) refers to the weighted average. To calculate the weighted average, each criterion's weight is multiplied by the alternative's score for that criterion, and the resulting value is divided by the sum of the weights:

$$\overline{X} = \frac{\sum_{i=0}^{n} W_i X_i}{\sum_{i=0}^{n} W_i}.$$

In multi-criteria decision-making methods, since the sum of criteria weights equals one, the denominator in the above formula becomes one and therefore does not need to be written. For calculating the weighted average in Excel, the SUMPRODUCT function is used.

Example 5.2 The performance metrics for five product models are provided in the table below. Apply the weighted product model (WPM) to rank these models and determine the most optimal one (Tables 5.5, 5.6 and 5.7).

In the weighted product model (WPM), the score for each alternative is computed as the product of the elements in each row of the weighted normalized decision matrix. The **PRODUCT** function in Excel computes the product of a specified range of values (Table 5.8).

Table 5.7 Weighted normalized decision matrix

	A	B	C	D	E	F
14		C_1^+	C_2^-	C_3^-	C_4^+	C_5^-
15	A1	= B9^\$B\$7 0.8034	=C9^\$C\$7 0.9499	=D9^\$D\$7 0.8409	0.9271	0.9585
16	A2	1.0000	1.0000	0.7598	0.9168	0.8572
17	A3	0.7433	0.9510	0.8409	0.9721	0.9585
18	A4	0.8034	0.9637	1.0000	0.9141	0.8766
19	A5	0.9344	0.9804	0.7311	1.0000	1.0000

Table 5.8 Weighted alternative ranking

	A	B	C
20		Ui	RANK
21	A1	=PRODUCT(B15:F15) 0.5703	=RANK(B21,\$B\$21:\$B\$25,0) 4
22	A2	=PRODUCT(B16:F16) 0.5703	3
23	A3	0.5539	5
24	A4	0.6204	2
25	A5	0.6697	1

5.4 Weighted Aggregated Sum Product Assessment (WASPAS)

The weighted aggregated sum product assessment (WASPAS) method is a modern multi-criteria decision-making (MCDM) technique developed by Zavadskas et al. (2012). This method integrates two well-established models: the weighted sum model (WSM) and the weighted product model (WPM), resulting in improved decision accuracy compared to using either model independently.

Like other MCDM techniques—such as ARAS, COPRAS, TOPSIS, and VIKOR—WASPAS operates on a decision matrix composed of criteria and alternatives. To rank alternatives effectively, the method requires criteria weights, which can be derived through various weighting approaches, including:

- Analytic hierarchy process (AHP)
- Shannon entropy
- Grey entropy
- Best-worst method (BWM)

Due to its hybrid structure, WASPAS has been widely applied in various fields, demonstrating robustness in ranking alternatives under multiple conflicting criteria.

Table 5.9 Decision matrix

	A	B	C	D	E
1		C_1^-	C_2^+	C_3^-	C_4^+
2	A_1	11	1.63	88	1.98
3	A_2	7	1.72	12.6	3.21
4	A_3	10	1.87	94	2.94
5	A_4	19	1.91	23	4.37
6	W_j	0.15	0.3	0.4	0.15

Step-by-Step WASPAS Procedure with Mathematical Formulas
Step 1: Construct the decision matrix (X).
Step 2: Normalize the decision matrix.
Step 3: Apply weighted sum model (WSM)—Q_1.

Calculate the weighted sum score for each alternative:

$$Q_{1i} = \sum_{j=1}^{n} w_j \times r_{ij}$$

where w_j is weight of criterion C_j (with $\sum W_j = 1$).

Step 4: Apply weighted product model (WPM)—Q_2.

Calculate the weighted product score for each alternative:

$$Q_{2i} = \prod_{j=1}^{n} \left(r_{ij}\right)^{w_j}.$$

Step 5: Compute the aggregated WASPAS score (Q).

Combine WSM and WPM using a weighting parameter (λ), typically set to $\lambda = 0.5$ for equal contribution:

$$Q_i = \lambda Q_{1i} + (1 - \lambda) Q_{2i}.$$

Step 6: Rank alternatives.

Rank alternatives based on descending values of Q (higher Q = better alternative).

***Example 5.3** Apply the weighted aggregated sum product assessment (WASPAS) method to determine the most suitable raw materials (A_1 through A_4) considering four key selection criteria (Tables 5.9, 5.10, 5.11 and 5.12).*

The PRODUCT function in Excel computes the multiplicative product of specified numerical values or a defined range of cells (Table 5.13).

Table 5.10 Normalized decision matrix

	A	B	C	D	E
7		C_1^-	C_2^+	C_3^-	C_4^+
8	A_1	=MIN(B$2:B$5)/B2 0.6364	=C2/MAX(C$2:C$5) 0.8534	0.1432	0.4531
9	A_2	=MIN(B$2:B$5)/B3 1.0000	0.9005	1.0000	0.7346
10	A_3	0.7000	0.9791	0.1340	0.6728
11	A_4	0.3684	1.0000	0.5478	1.0000

Table 5.11 Weighted normalized decision matrix

	A	B	C	D	E	F
12		C_1^-	C_2^+	C_3^-	C_4^+	Q_{1i}
13	A1	=B8*B$6 0.0955	=C9*C$6 0.2560	0.0573	0.0680	=SUM(B13:E13) 0.4767
14	A2	=B9*B$6 0.1500	0.2702	0.4000	0.1102	=SUM(B14:E14) 0.9303
15	A3	0.1050	0.2937	0.0536	0.1009	0.5532
16	A4	0.0553	0.3000	0.2191	0.1500	0.7244

Table 5.12 Squired normalized decision matrix

	A	B	C	D	E	F
17		C_1^-	C_2^+	C_3^-	C_4^+	Q_{2i}
18	A1	=B8^B$6 0.9344	=C8^C$6 0.9536	0.4596	0.8880	=PRODUCT(B18:E18) 0.3637
19	A2	1.0000	0.9691	1.0000	0.9548	0.9252
20	A3	0.9479	0.9937	0.4476	0.9423	0.3973
21	A4	0.8609	1.0000	0.7861	1.0000	0.6767

Table 5.13 Weighted alternative ranking

	A	B
22	Q_i	RANK
23	=SUM(F13+F18)/2 0.4202	=RANK(A23,A$23:A$26) 4
24	0.9278	1
25	0.4753	3
26	0.7006	2

5.5 Multiple Attribute Utility Theory (MAUT)

The MAUT method focuses on constructing a decision-maker's (DM) utility function for each individual criterion and then aggregating these functions to evaluate the overall utility of each alternative. This technique, introduced into the multi-criteria decision-making framework by Keeney and Raiffa in 1976, aims to rank decision alternatives based on predefined criteria.

The concept of "utility" reflects the DM's preference for specific outcomes. Given a predefined set of performance criteria, a mathematical utility function can be formulated for each criterion, representing the DM's satisfaction level across all possible outputs. In decision theory, such a function is termed a "utility function." Utility captures the maximum attainable satisfaction a DM associates with a given objective (or criterion). By solving the decision problem through utility function aggregation, an optimal solution that maximizes the DM's satisfaction is derived.

Step-by-Step MAUT Implementation for Optimal Alternative Selection

1. **Identifying Criteria and Alternatives**

   - Define decision criteria $(C_1, C_2, \ldots, C_n)$.
   - Determine alternatives $(A_1, A_2, \ldots, A_m)$.

2. **Normalizing Criterion Values**

   For positive criteria (higher is better):

$$x_{ij}^* = \frac{x_{ij} - \min(x_j)}{\max(x_j) - \min(x_j)}.$$

   For negative criteria (lower is better):

$$x_{ij}^* = \frac{\max(x_{ij}) - x_{ij}}{\max(x_j) - \min(x_j)}.$$

3. **Defining Utility Functions for Each Criterion**

   Linear utility function:

$$U_j\left(x_{ij}^*\right) = x_{ij}^*.$$

   Nonlinear (e.g., exponential) utility function (if needed):

$$U_j\left(x_{ij}^*\right) = 1 - e^{-kx_{ij}^*}.$$

   Table 5.14 presents the specification of utility functions corresponding to different risk attitudes for each decision criterion.

4. **Assigning Weights to Criteria** (W_j)
5. **Calculating Overall Utility for Each Alternative**

Table 5.14 Specification of utility functions

Criterion type	Risk type	The utility function (U_{ij})
Positive	Risk-averse	$u_{ij} = r_{ij}$
	Risk-seeking	$u_{ij} = r_{rj}^{k_j}$
	Extreme risk	$u_{ij} = \sqrt[k_j]{r_{ij}}$
Negative	Risk-averse	$u_{ij} = 1 - r_{ij}$
	Risk-seeking	$u_{ij} = 1 - r_{rj}^{k_j}$
	Extreme risk	$u_{ij} = \left(r_{ij} - 1\right)^{k_j}$

Table 5.15 Decision matrix

	A	B	C	D	E	F
1		C_1^+	C_2^+	C_3^-	C_4^+	C_5^+
2	A_1	2	180	5.4	0.75	20000
3	A_2	2.5	150	5.7	0.93	18000
4	A_3	1.4	180	3.8	0.90	20000
5	A_4	1.9	240	7.2	0.84	21000
6	A_5	2.5	200	6.2	0.88	27000
7	A_6	2.2	180	5.9	0.95	20000
8	W_j	0.15	0.3	0.3	0.13	0.12
9	Risk preference	Risk-neutral	Risk-seeking	Risk-seeking	Risk-neutral	Risk-averse
10	Upper bound (expert estimate)	4	300	8	1	30000
11	Lower bound (expert estimate)	0.5	100	2.5	0.5	18000

$$U(A_i) = \sum_{j=1}^{n} W_j U_j\left(x_{ij}^*\right).$$

6. **Ranking Alternatives**

Alternatives are ranked from highest to lowest based on $U(A_i)$.

Example 5.4 The dataset presented in Table 5.15 contains evaluations of six projects across five specified criteria, each with established risk parameters. Conduct project prioritization by applying the multi-attribute utility technique (MAUT) (Tables 5.16, 5.17 and 5.18).

Table 5.16 Normalized decision matrix

	A	B	C	D	E	F
12		C_1^+	C_2^+	C_3^-	C_4^+	C_5^+
13	A1	=(B2-B$11)/ (B$10-B$11) 0.4286	=(C2-C$11)/ (C$10-C$11) 0.4000	=(D2-D$11)/ (D$10-D$11) 0.5273	0.5000	0.1667
14	A2	=(B3-B$11)/ (B$10-B$11) 0.5714	0.2500	0.5818	0.8600	0.0000
15	A3	0.2571	0.4000	0.2364	0.8000	0.1667
16	A4	0.4000	0.7000	0.8545	0.6800	0.2500
17	A5	0.5714	0.5000	0.6727	0.7600	0.7500
18	A6	0.4857	0.4000	0.6182	0.9000	0.1667

Note: Normalization was performed irrespective of criterion polarity (positive/negative nature)

Table 5.17 Utility matrix

	A	B	C	D	E	F
19		C_1^+	C_2^+	C_3^-	C_4^+	C_5^+
20	A1	=B13 0.4286	K=3 =C13^3 0.064	=1-(D13^3) 0.8944	=E13 0.5000	K=3 =F13^(1/3) 0.5503
21	A2	=B14 0.5714	0.016	0.9269	0.8600	0.0000
22	A3	0.2571	0.064	0.5547	0.8000	0.5503
23	A4	0.4000	0.343	0.9969	0.6800	0.6300
24	A5	0.5714	0.125	0.9649	0.7600	0.9086
25	A6	0.4857	0.064	0.9443	0.9000	0.5503

Table 5.18 Utility matrix weighted

	A	B	C	D	E	F	G	H
26		C_1^+	C_2^+	C_3^-	C_4^+	C_5^+	Total U	RANK
27	A1	=B20*B$8 0.0643	=C20*C$8 0.0192	0.2560	0.0650	0.0660	=SUM(B27: F27) 0.4705	=RANK (G27, $G$27:$ G$32) 4
28	A2	0.0857	0.0047	0.2409	0.1118	0.0000	0.4431	5
29	A3	0.0386	0.0192	0.2960	0.1040	0.0660	0.5238	2
30	A4	0.0600	0.1029	0.1128	0.0884	0.0756	0.4397	6
31	A5	0.0857	0.0375	0.2087	0.0988	0.1090	0.5397	1
32	A6	0.0729	0.0192	0.2291	0.1170	0.0660	0.5042	3

5.6 The Complex Proportional Assessment Technique (COPRAS)

The COPRAS method, a prominent MCDM technique, ranks and selects the optimal alternative from a given set by evaluating each option's relative closeness to the ideal solution and its deviation from the worst-case scenario (Zavadskas & Kaklauskas, 1996).

COPRAS Implementation Steps

1. Decision matrix formation
2. Normalization

$$Z_{ij} = \frac{X_{ij}}{\sum_{i=1}^{m} X_{ij}}$$

3. Criteria weighting
4. Weighted normalized matrix
5. Separation of beneficial and non-beneficial criteria
6. Relative utility (Q_i) calculation
7. Final ranking

Example 5.5 Apply the COPRAS method to prioritize the alternatives based on five predefined criteria (Tables 5.19, 5.20, 5.21 and 5.22).

Table 5.19 Decision matrix

	A	B	C	D	E	F
1		C_1^+	C_2^-	C_3^+	C_4^-	C_5^-
2	A1	2	46	21000	210	4000
3	A2	1	28	14000	270	3500
4	A3	5	63	6500	190	3800
5	A4	3	25	13000	400	4100
6	Wj	0.12	0.17	0.28	0.2	0.23

Table 5.20 Normalized decision matrix

	A	B	C	D	E	F
7		C_1^+	C_2^-	C_3^+	C_4^-	C_5^-
8	A1	=B2/SUM(B$2:B$5) 0.1818	=C2/SUM(C$2:C$5) 0.2840	0.3853	0.1963	0.2597
9	A2	0.0909	0.1728	0.2569	0.2523	0.2273
10	A3	0.4545	0.3889	0.1193	0.1776	0.2468
11	A4	0.2727	0.1543	0.2385	0.3738	0.2662

Table 5.21 Weighted normalized decision matrix

	A	B	C	D	E	F	G	H
12		C_1^+	C_2^-	C_3^+	C_4^-	C_5^-	s^+	s^-
13	A1	=B8*B$6 0.0218	=C8*C$6 0.0483	0.1079	0.0393	0.0597	=B13+D13 0.1297	=C13+E13+F13 0.1473
14	A2	0.0109	0.0294	0.0719	0.0505	0.0523	0.0828	0.1321
15	A3	0.0545	0.0661	0.0334	0.0355	0.0568	0.0879	0.1584
16	A4	0.0327	0.0262	0.0668	0.0748	0.0612	0.0995	0.1622

Table 5.22 Final ranking

	A	B	C	D
17		$\frac{1}{s^-}$	U	R
18	A1	=1/ H13 6.7905	=G13+((SUM(H$13:H$16))/(H13*SUM(B$18: B$21))) 0.2815	=RANK(C18, C$18:C$21) 1
19	A2	7.5687	0.2521	2
20	A3	6.3140	0.2291	4
21	A4	6.1639	0.2373	3

References

Bridgman, P. W. (1922). *Dimensional analysis*. Yale University Press.

Hwang, C. L., & Yoon, K. (1981). *Multiple attribute decision making: Methods and applications* (Lecture Notes in Economics and Mathematical Systems) (Vol. 186). Springer.

Keeney, R. L., & Raiffa, H. (1976). *Decisions with multiple objectives: Preferences and value tradeoffs*. Wiley.

Miller, D. W., & Starr, M. K. (1960). *Executive decisions and operations research*. Prentice-Hall.

Miller, D. W., & Starr, M. K. (1969). *Executive decisions and operations research*. Prentice-Hall.

Zavadskas, E. K., & Kaklauskas, A. (1996). *Multiple criteria evaluation of buildings*. Technika.

Zavadskas, E. K., Turskis, Z., Antucheviciene, J., & Zakarevicius, A. (2012). Optimization of weighted aggregated sum product assessment. *Elektronika ir Elektrotechnika, 122*(6), 3–6. https://doi.org/10.5755/j01.eee.122.6.1810

Glossary

Analytic Hierarchy Process (AHP) A structured MCDM method that decomposes a complex decision problem into a hierarchy of goal, criteria, and alternatives. It uses pairwise comparisons to derive ratio scales and prioritize alternatives.

Analytic Network Process (ANP) A generalization of AHP that models complex decision problems as a network, allowing for interdependencies and feedback relationships between decision elements (clusters and nodes), rather than a simple hierarchy.

Best-Worst Method (BWM) An MCDM method where the decision-maker identifies the best and worst criteria and then performs pairwise comparisons only between these and the other criteria, reducing the number of comparisons and improving consistency.

Chebyshev Norm A norm that calculates the maximum absolute difference between the elements of two vectors. In MCDM, it is used when the focus is on the worst deviation across criteria.

Compromise Solution A solution found by methods like VIKOR that balances the objectives of maximizing group utility and minimizing individual regret, representing a feasible alternative that is closest to the ideal.

Concordance Set In ELECTRE methods, the set of criteria for which an alternative A_l is at least as good as another alternative A_h.

Consistency Index (CI) A measure of the deviation from perfect consistency in a pairwise comparison matrix.

Consistency Ratio (CR) A ratio calculated by comparing the Consistency Index (CI) to a Random Index (RI). A CR of 0.1 or less is generally considered acceptable, indicating that the judgments are sufficiently consistent.

Constraint Limitations or restrictions imposed on the decision variables, defining the boundaries of the feasible decision space.

Content Validity Index (CVI) A measure of content validity, calculated as the proportion of experts who rate an item as relevant (a score of 3 or 4 on a four-point scale). The minimum acceptable value is typically 0.79.

A. Amirteimoori, M. Soufi, *Industrial Decision Analysis*, Multiple Criteria Decision Making, https://doi.org/10.1007/978-3-032-24299-0

Content Validity Ratio (CVR) A measure of content validity developed by Lawshe, used to assess whether an item is essential for the operationalization of a construct. Its minimum acceptable value depends on the number of experts.

COPRAS (Complex Proportional Assessment) An MCDM method that evaluates alternatives by their relative closeness to the ideal solution and their deviation from the worst-case scenario. It explicitly separates beneficial and non-beneficial criteria in its calculation.

DANP (DEMATEL-based ANP) A hybrid MCDM method that combines DEMATEL and ANP. DEMATEL is first used to map the interdependencies and influence relationships between criteria, which then informs the network structure for the ANP analysis.

Decision Matrix A table where rows represent alternatives and columns represent criteria, with each cell (x_{ij}) showing the performance value of alternative i with respect to criterion j.

Decision Space The set of all possible decisions or values that the decision variables can take, subject to the problem's constraints.

Decision Variable A variable whose value can be controlled or chosen by the decision-maker to achieve the desired objectives.

Decision Making Trial and Evaluation Laboratory (DEMATEL) A method for constructing and analyzing a structural model of causal relationships between complex factors. It visualizes these relationships and identifies which factors are net causes and which are net effects.

Discordance Set In ELECTRE methods, the set of criteria for which an alternative A_l is not at least as good as another alternative A_h (i.e., A_h is strictly better).

Dominant Solution A feasible solution that is at least as good as another feasible solution in all objectives and strictly better in at least one objective.

Eigenvalue Method A primary technique for deriving the priority vector (weights) from a pairwise comparison matrix in AHP, based on calculating the principal eigenvector of the matrix.

ELECTRE (ELimination and Choice Translating REality) A family of outranking MCDM methods that compare alternatives pairwise to build outranking relations, which are then used to eliminate less favorable alternatives. Different versions (I, II, III, TRI) are used for selection, ranking, or classification problems.

Epsilon-Constraint Method A technique for solving MODM problems where one objective is optimized while all other objectives are converted into constraints with specified bounds.

Euclidean Norm (L2-norm) A norm that calculates the straight-line distance between two points in Euclidean space. It is the most common distance measure used in TOPSIS.

Extended TOPSIS A version of TOPSIS designed for group decision-making. It involves calculating distances for each decision-maker separately and then aggregating them using the geometric mean.

Feasible Solution A set of values for the decision variables that satisfies all the constraints of the problem.

Fuzzy Logic A multi-valued logic system that deals with reasoning that is approximate rather than fixed and exact. It uses membership functions with values between 0 and 1 to handle vagueness and uncertainty, often used to translate linguistic terms (e.g., "high impact") into fuzzy numbers for analysis.

Fuzzy Numbers A mathematical representation of a fuzzy quantity, such as "approximately 5," often depicted as triangular (TFN) or trapezoidal fuzzy numbers with a range of possible values and a membership function.

Fuzzy TOPSIS An extension of the TOPSIS method that uses fuzzy numbers to represent the performance of alternatives and the weights of criteria, allowing it to handle uncertainty and imprecision in human judgments.

Geometric Mean A type of average calculated by multiplying a set of numbers and then taking the *n*th root. It is the preferred method for aggregating the judgments of multiple experts in AHP and other pairwise comparison methods.

Goal (or Target) A specific, desired level of achievement for an objective. It adds a quantifiable target to an objective.

Goal Programming A popular MODM technique that assigns target values (goals) to each objective and seeks to minimize the deviations from these goals.

Group Utility (S) In VIKOR, a measure that represents the total weighted distance of an alternative from the ideal solution, reflecting the "majority" or overall performance.

Hamming Distance A norm that calculates the sum of the absolute differences between the elements of two vectors. It is used as an alternative distance measure in improved TOPSIS.

Ideal Solution (Positive Ideal Solution) A hypothetical alternative that represents the best possible performance on all criteria (the maximum value for benefit criteria and the minimum for cost criteria).

Improved TOPSIS A variant of TOPSIS that combines multiple distance measures (e.g., Euclidean, Hamming, Grey) to calculate a composite score for ranking alternatives, aiming for more robust results.

Incomparability A state in outranking methods where, based on the available information, it is not possible to establish a preference or indifference relationship between two alternatives.

Indifference A state where two alternatives are considered equal in value or preference from the decision-maker's perspective.

Indifference Threshold (q) In outranking methods, the largest difference between two alternatives on a criterion that is considered negligible by the decision-maker, leading to a state of indifference.

Individual Regret (R) In VIKOR, a measure that represents the maximum weighted distance of an alternative from the ideal solution for any single criterion, reflecting the "opponent's" perspective or the worst-case performance.

Least Squares Method (LSM) An optimization technique used to find the best fit for a model by minimizing the sum of the squares of the differences between observed and predicted values. In MCDM, it's used to find weights that minimize inconsistency in pairwise comparison matrices.

Linguistic Variable A variable whose values are not numbers but words or sentences in a natural or artificial language. In fuzzy MCDM, these are used to capture subjective expert judgments (e.g., "Strong Preference," "Moderate Impact").

LINMAP (Linear Programming Technique for Multidimensional Analysis of Preference) An MCDM method that uses linear programming to analyze pairwise preferences provided by a decision-maker. It aims to find an ideal point and weights for criteria that best satisfy these preferences, minimizing the sum of inconsistency errors.

Logarithmic Fuzzy Priority Programming (LFPP) An advanced MCDM technique that combines fuzzy logic, logarithmic transformations, and priority programming. It is designed to handle uncertainty and nonlinearity in determining criteria weights and prioritizing alternatives from fuzzy pairwise comparison matrices.

Logarithmic Least Squares Method (LLSM) A variant of the Least Squares Method applied to logarithmic transformations of data. In MCDM, particularly within AHP, it's a popular method for deriving priority vectors from pairwise comparison matrices, as its solution is equivalent to the geometric mean method for consistent matrices.

Logarithmic Priority Programming (LPP) A mathematical optimization technique that uses logarithmic functions to transform and normalize criteria values, combined with priority programming to determine optimal weights and priorities, especially useful for handling nonlinear relationships in data.

Mahalanobis Distance A distance measure that accounts for the correlations between variables (criteria). It is used in Improved TOPSIS when criteria are not independent.

Manhattan Norm (L1-norm) A norm that calculates the distance between two vectors as the sum of the absolute differences of their coordinates. It is used in methods like VIKOR.

MAUT (Multiple Attribute Utility Theory) A decision-making technique that constructs a utility function for each criterion to represent a decision-maker's preference or satisfaction level. The overall utility of an alternative is the weighted sum of these individual utilities.

Multi-Objective Decision Making (MODM) A field of management science and operations research that deals with decision-making problems characterized by multiple, often conflicting, objectives that need to be optimized simultaneously.

Negative Ideal Solution (NIS) A hypothetical alternative that represents the worst possible performance on all criteria (the minimum value for benefit criteria and the maximum for cost criteria).

Net Effect (R-C) In DEMATEL, the difference between the total influence a factor exerts (R) and the total influence it receives (C). A positive value indicates a net cause, while a negative value indicates a net effect.

Net Flow In PROMETHEE II, the difference between the positive outranking flow (Φ+Φ+) and the negative outranking flow (Φ−Φ−) for an alternative, used to derive a complete ranking.

Non-dominated Solution See **Pareto Optimal Solution**.

Norm A mathematical function that assigns a non-negative length or size to a vector. In MCDM, norms are used to measure distances between alternatives and reference points.

Normalization The process of transforming data from different scales to a common, dimensionless scale, allowing for fair comparison and aggregation. Common methods include sum normalization, vector normalization, and linear max-min normalization.

Objective A specific aim or goal that a decision-maker wants to achieve, often expressed as a function of the decision variables.

Objective Space The set of all possible combinations of objective function values that can be achieved by feasible solutions.

Optimal Solution In single-objective problems, the feasible solution that provides the best value for the objective. In MODM, this term often refers to an *ideal* solution that optimizes all objectives, which is generally unattainable.

Outranking Relation A binary relation used in ELECTRE and PROMETHEE, where A_l outranks A_h if there is sufficient evidence to support the claim that A_l is at least as good as A_h.

Pairwise Comparison The fundamental process in many MCDM methods where two elements (criteria or alternatives) are compared against each other at a time to determine a preference, using a defined scale like Saaty's 1–9 scale.

Pareto Front The set of all Pareto optimal solutions, plotted in the objective space. It represents the trade-off frontier.

Pareto Optimal Solution A solution where no other feasible solution can improve at least one objective without worsening at least one other objective. Also known as a non-dominated solution.

Payoff Matrix A table used in interactive methods like STEM, showing the performance of each objective function at the optimal solution of each individual objective.

Preference Data Information provided by a decision-maker in the form of pairwise comparisons (e.g., "alternative A is preferred over alternative B"), which is used as input for methods like LINMAP.

Preference Function In PROMETHEE, a function that translates the difference between two alternatives on a given criterion into a preference degree ranging from 0 to 1. Six types are defined to suit different data characteristics.

Preference Threshold (p) In outranking methods, the smallest difference between two alternatives on a criterion that is sufficient to generate a strict preference for the better one.

Preferred Solution A non-dominated solution that is ultimately chosen by the decision-maker as the final solution based on their unique preferences and other qualitative criteria.

Prominence (R+C) In DEMATEL, the sum of the total influence a factor exerts (R) and receives (C). It indicates the total strength of influence of that factor within the system, or its overall importance.

PROMETHEE (Preference Ranking Organization METHod for Enrichment Evaluations) A family of outranking MCDM methods that use preference functions to model the decision-maker's preference for one alternative over another on each criterion.

PROMETHEE I Provides a partial ranking of alternatives.

PROMETHEE II Provides a complete ranking of alternatives based on net flow.

Pseudo-Criterion A type of criterion used in outranking methods that introduces thresholds (indifference and preference) to account for the imprecision and uncertainty in data and preferences. This contrasts with a "true criterion" where any difference leads to a preference.

Random Index (RI) A pre-calculated average Consistency Index (CI) for randomly generated pairwise comparison matrices of a given size (n). It is used in the calculation of the Consistency Ratio (CR).

Relative Closeness (C_i) In TOPSIS, an index that measures how close an alternative is to the ideal solution and how far it is from the negative-ideal solution. Alternatives are ranked by this value in descending order.

Saaty's Scale A 1–9 ratio scale used in AHP and other pairwise comparison methods to quantify qualitative judgments, where 1 represents equal importance and 9 represents extreme importance of one element over another.

Satisfactory Solution A solution that meets the decision-maker's desired levels for each of the objectives.

SAW (Simple Additive Weighting) Also known as the Weighted Sum Model (WSM). A fundamental MCDM method that calculates a total score for each alternative as the weighted sum of its normalized performance values. Alternatives are then ranked by this score.

STEM Method (Step Method) One of the first interactive approaches for solving multi-objective optimization problems, where the decision-maker provides preference information iteratively to converge on a satisfactory solution.

Supermatrix In ANP, a large partitioned matrix that represents the influence of every element in the network on every other element. It is built from the priority vectors derived from pairwise comparisons.

Total Relation Matrix (TRM) In DEMATEL, a matrix that captures both the direct and indirect influences between factors. It is derived from the normalized direct relation matrix and is used to calculate the prominence and net effect of each factor.

TOPSIS (Technique for Order of Preference by Similarity to Ideal Solution) See entry under 'T'.

Trade-off Improving one objective at the expense of another.

VIKOR (VlseKriterijumska Optimizacija I Kompromisno Resenje) A distance-based MCDM method that focuses on ranking and selecting from a set of alternatives in the presence of conflicting criteria. It introduces a compromise solution based on the measures of group utility (S) and individual regret (R).

WASPAS (Weighted Aggregated Sum Product Assessment) A hybrid MCDM method that combines the results from the Weighted Sum Model (WSM/SAW) and the Weighted Product Model (WPM) to achieve higher ranking accuracy. The final score is a weighted combination of the two.

Weight A numerical value assigned to an objective or criterion to represent its relative importance compared to others in the decision-making process.

Weighted Norm A norm where each element of the vector is multiplied by a weight before the distance is calculated, reflecting the relative importance of different criteria.

Weighted Normalized Matrix A matrix obtained by multiplying each element of the normalized decision matrix by the weight of its corresponding criterion.

Weighted Product Model (WPM) An MCDM method where the overall score for an alternative is calculated by raising each normalized performance value to the power of its criterion's weight and then taking the product of these results.

Weighted Sum Method A technique that combines multiple objectives into a single objective function by multiplying each objective by a user-specified weight and then summing them. See also **SAW**.

Win Rate Method A simple method for calculating pairwise comparison results by tallying the number of times each alternative "wins" in head-to-head comparisons and expressing it as a percentage.

Index

A. Amirteimoori, M. Soufi, *Industrial Decision Analysis*, Multiple Criteria Decision Making, https://doi.org/10.1007/978-3-032-24299-0

MIX
Papier aus verantwortungsvollen Quellen
Paper from responsible sources
FSC® C105338

If you have any concerns about our products,
you can contact us on
ProductSafety@springernature.com

In case Publisher is established outside the EU,
the EU authorized representative is:
Springer Nature Customer Service Center GmbH
Europaplatz 3, 69115 Heidelberg, Germany

Printed by Libri Plureos GmbH
in Hamburg, Germany